閻老師指正：

生 龍溪敬贈

MICROBIAL PROCESSES:
Promising Technologies for Developing Countries

Report of an Ad Hoc Panel of the
Advisory Committee on Technology Innovation
Board on Science and Technology for International Development
Commission on International Relations
National Research Council

NATIONAL ACADEMY OF SCIENCES
Washington, D.C. 1979

This report has been prepared by an ad hoc advisory panel of the Board on Science and Technology for International Development, Commission on International Relations, National Research Council, for the Office of Science and Technology, Bureau for Technical Assistance, Agency for International Development, Washington, D.C., under Contract No. AID/csd-2584, Task Order No. 1.

NOTICE: The project that is the subject of this report was approved by the Governing Board of the National Research Council, whose members are drawn from the Councils of the National Academy of Sciences, the National Academy of Engineering, and the Institute of Medicine. The members of the Committee responsible for the report were chosen for their special competences and with regard for appropriate balance.

This report has been reviewed by a group other than the authors according to procedures approved by a Report Review Committee consisting of members of the National Academy of Sciences, the National Academy of Engineering, and the Institute of Medicine.

Library of Congress Catalog Number 79-91534

Panel on Microbial Processes

J. ROGER PORTER, Department of Microbiology, University of Iowa, Iowa City, Iowa, *Chairman*

ROBERT F. ACKER, American Society for Microbiology, Washington, D.C.

ARTHUR W. ANDERSON, Department of Microbiology, Oregon State University, Corvallis, Oregon

WINTHROP D. BELLAMY, Department of Food Science, Cornell University, Ithaca, New York

WAYNE M. BROOKS, Department of Entomology, University of North Carolina, Raleigh, North Carolina

MARVIN P. BRYANT, Department of Dairy Science, College of Agriculture, University of Illinois, Urbana, Illinois

LEE A. BULLA, JR., U.S. Grain Marketing Research Laboratory, Federal Research, North Central Region, Science and Education Administration, U.S. Department of Agriculture, Manhattan, Kansas

JOE C. BURTON, NITRAGIN, Milwaukee, Wisconsin

STAFFAN DELIN, Berkeley, California

RICHARD D. DONOVICK, American Type Culture Collection, Rockville, Maryland

EUGENE L. DULANEY, Merck Institute for Therapeutic Research, Rahway, New Jersey

LLOYD R. FREDERICK (formerly, Department of Agronomy, Iowa State University, Ames, Iowa), Senior Microbiologist, Office of Agriculture, Tropical Soil and Water Management Division, Development Support Bureau, Agency for International Development, Washington, D.C.

JAMES W. GERDEMANN, Department of Plant Pathology, University of Illinois, Urbana, Illinois

CLARENCE G. GOLUEKE, Cal Recovery Systems Incorporated, Richmond, California

RICHARD S. HANSON, Department of Bacteriology, University of Wisconsin, Madison, Wisconsin

CLIFFORD W. HESSELTINE, Northern Regional Research Laboratory, U.S. Department of Agriculture, Peoria, Illinois

GLADYS L. HOBBY (formerly, Chief, Special Research Laboratory [Infectious Diseases], U. S. Veterans Administration, Cornell University Medical College, New York, New York), Editor-in-Chief, Antimicrobial Agents and Chemotherapy, Kennett Square, Pennsylvania

RILEY D. HOUSEWRIGHT, Committee on Toxicology, National Academy of Sciences, Washington, D.C.

CARLO M. IGNOFFO, Biological Control of Insects Research Laboratory, Entomology Resources Division, U.S. Department of Agriculture, Science and Education Administration, Research Park, Columbia, Missouri

T. KENT KIRK, U.S. Forest Products Laboratory, U.S. Department of Agriculture, Forest Service, Madison, Wisconsin

ALLEN I. LASKIN, EXXON Research and Engineering Company, Linden, New Jersey

JOHN H. LITCHFIELD, Battelle Memorial Institute, Columbus Laboratories, Columbus, Ohio

DONALD H. MARX, Forestry Science Laboratory, U.S. Department of Agriculture, Forest Service, Athens, Georgia

WILLIAM J. OSWALD, Professor of Sanitary Engineering and Public Health, University of California, Berkeley, California

BURTON M. POGELL, Department of Microbiology, School of Medicine, St. Louis University, St. Louis, Missouri

DAVID PRAMER, Associate Vice President for Research, Rutgers University, New Brunswick, New Jersey

DONALD W. ROBERTS, Boyce Thompson Institute for Plant Research, Cornell University, Ithaca, New York

OLDRICH K. SEBEK, Infectious Diseases Research, The Upjohn Company, Kalamazoo, Michigan

KEITH H. STEINKRAUS, Department of Food Science and Technology, New York State Agricultural Experiment Station, Geneva, New York

DONALD K. WALTER, Urban Waste Technology, U.S. Department of Energy, Washington, D.C.

DEANE F. WEBER, Cell Culture and Nitrogen Fixation Laboratory, U.S. Department of Agriculture, Beltsville, Maryland

BERNARD A. WEINER, Northern Regional Research Center, U.S. Department of Agriculture, Peoria, Illinois

WILLIAM E. WOODWARD, Program in Infectious Diseases and Clinical Microbiology, University of Texas Health Science Center, Houston, Texas

OSKAR R. ZABORSKY, National Science Foundation, Washington, D.C.

NAS Steering Committee

J. ROGER PORTER, Department of Microbiology, University of Iowa, Iowa City, Iowa

ROBERT F. ACKER, American Society for Microbiology, Washington, D.C.

RILEY D. HOUSEWRIGHT, Committee on Toxicology, National Academy of Sciences, Washington, D.C.

Study Staff

B. K. WESLEY COPELAND, Board on Science and Technology for International Development, Commission on International Relations, National Academy of Sciences–National Research Council, Washington, D.C.

M. G. C. McDONALD DOW, Board on Science and Technology for International Development, Commission on International Relations, National Academy of Sciences–National Research Council, Washington, D.C.

E. GRIFFIN SHAY, Board on Science and Technology for International Development, Commission on International Relations, National Academy of Sciences–National Research Council, Washington, D.C.

Contributors

MARTIN ALEXANDER, Department of Soil Science, Cornell University, Ithaca, New York

LARRY L. ANDERSON, Department of Mining & Fuels Engineering, University of Utah, Salt Lake City, Utah

DURWARD BATEMAN, President, Plant Pathology Society, Cornell University, Ithaca, New York

L. JOE BERRY, Department of Microbiology, University of Texas, Austin, Texas

JOHN J. BOLAND, Department of Geography and Environmental Engineering, The Johns Hopkins University, Baltimore, Maryland

HENRY R. BUNGAY, Department of Chemical and Environmental Engineering, Rensselaer Polytechnic Institute, Troy, New York

ROBERT H. BURRIS, Department of Biochemistry, University of Wisconsin, Madison, Wisconsin

R. R. COLWELL, Director, Sea Grant Program, University of Maryland, College Park, Maryland

R. JAMES COOK, Regional Cereal Disease Research Laboratory, USDA, Washington State University, Pullman, Washington

THOMAS M. COOK, Department of Microbiology, University of Maryland, College Park, Maryland

DON L. CRAWFORD, Department of Bacteriology & Biochemistry, College of Agriculture, University of Idaho, Moscow, Idaho

CONSTANT C. DELWICHE, Department of Land, Air and Water Resources, University of California, Davis, California

RAYMOND N. DOETSCH, Department of Microbiology, University of Maryland, College Park, Maryland

LOUIS A. FALCON, Department of Entomology, University of California, Berkeley, California

RICHARD A. FINKELSTEIN, Department of Microbiology, University of Texas, Dallas, Texas

E. M. FOSTER, Director, Food Research Institute, University of Wisconsin, Madison, Wisconsin

HARLYN O. HALVORSON, Director, Rosenstiel Basic Medical Sciences Research Center, Brandeis University, Waltham, Massachusetts

ROBERT P. HANSON, Department of Veterinary Science, University of Wisconsin, Madison, Wisconsin

CARL-GÖRAN HEDÉN, Karolinska Institute, Stockholm, Sweden

DAVID HENDLIN, Senior Director, Developmental Microbiology, Merck Sharp & Dohme Laboratories, Rahway, New Jersey

H. HEUKELEKIAN, New York, New York

WILLIAM N. HUBBARD, JR., President, The Upjohn Company, Kalamazoo, Michigan

ARTHUR E. HUMPHREY, Dean, College of Engineering and Applied Science, University of Pennsylvania, Philadelphia, Pennsylvania

J. W. M. LA RIVIERE, International Institute for Hydraulic and Environmental Engineering, Delft, The Netherlands

RAYMOND C. LOEHR, Director, Environmental Studies Program, Riley-Robb Hall, Cornell University, Ithaca, New York

CLAYTON W. McCOY, Institute of Food and Agricultural Sciences, University of Florida, Lake Alfred, Florida

WALSH McDERMOTT, Robert Wood Johnson Foundation, Princeton, New Jersey

ROSS E. McKINNEY, N. T. Veatch Professor of Environmental Health, Department of Civil Engineering, University of Kansas, Lawrence, Kansas

ROBERT A. MAH, Division of Environmental & Nutritional Studies, University of California School of Public Health, Los Angeles, California

STAN M. MARTIN, National Research Council of Canada, Ottawa, Canada

EMIL M. MRAK, Chancellor *Emeritus*, University of California, Davis, California

DANIEL J. O'NEIL, Engineering Experiment Station, Georgia Institute of Technology, Atlanta, Georgia

H. PEPPLER, White Fish Bay, Wisconsin

ELWYN T. REESE, Food Sciences Laboratory, U.S. Army Natick Research and Development Command, Natick, Massachusetts

MARTIN H. ROGOFF, Senior Staff Scientist, Hazard Evaluation Division, Office of Pesticide Programs, U.S. Environmental Protection Agency, Washington, D.C.

JAMES P. SAN ANTONIO, Science and Education Administration, U.S. Department of Agriculture, Beltsville, Maryland

WILLIAM D. SAWYER, Chairman, Department of Microbiology and Immunology, School of Medicine, Indiana University, Indianapolis, Indiana

RICHARD SOPER, Acting Research Leader, Insect Pathology Research Institute, Boyce Thompson Institute, U.S. Department of Agriculture, Ithaca, New York

KENNETH V. THIMANN, Thimann Laboratories, University of California, Santa Cruz, California

GEORGE T. TSAO, Laboratory of Renewable Resources, Purdue University, West Lafayette, Indiana

D. M. UPDEGRAFF, Chemistry and Geochemistry Department, Colorado School of Mines, Golden, Colorado

ABEL WOLMAN, The Johns Hopkins University, Baltimore, Maryland

B. C. WOLVERTON, National Space Technology Laboratories, NSTL Station, Mississippi

LUNG-CHI WU, Campbell Institute for Agricultural Research, Napoleon, Ohio

Roger Porter, who directed the organization and preparation of this report, died on May 24, 1979. Dr. Porter will be remembered for his unflagging dedication to the use of science for the benefit of mankind and for his warm and gracious manner in pursuing this purpose.

Preface

The National Academy of Sciences–National Research Council, through the Commission on International Relations and its Board on Science and Technology for International Development (BOSTID), has investigated scientific and technological advances that may be applicable to the less-developed regions of the world. BOSTID's Advisory Committee on Technology Innovation (ACTI) has reviewed a number of technologies in the United States and elsewhere to assess their potential for contributing to the economic and social well-being of those in developing countries. It is in this context that this study of microbiological processes was carried out.

An ad hoc panel of ACTI convened in August 1977 to select a group of microbial processes with promise for wider use in the developing world. To make the selection process more manageable, a steering committee of the panel chose ten subject areas they felt were most important to developing countries. In each subject area, specialists from the panel were asked to analyze the responses from a questionnaire sent to approximately 25,000 biological scientists and engineers. Each subpanel selected a small number of examples of microbial processes that met either of the following criteria:

- The process can be beneficially employed in developing countries

or

- The process has sufficient potential for developing countries to merit research and development for future use.

Because of the unique conditions in each country where the processes may be used, no attempt has been made to quantify economic feasibility. Depending on indigenous needs and resources, a process appropriate for one country may be inappropriate for another.

Assistance in reaching technical or economic conclusions concerning the various processes may be solicited from the individuals and institutions cited throughout the report.

For the convenience of the reader, each process is presented in a separate chapter, giving the following information:

- Methodology
- Potential value of the process
- Special needs and limitations
- Research and development requisites
- Suggested readings
- Sources for obtaining microorganisms.

In addition, the Introduction provides a nontechnical summary of the processes described in each chapter and characterizes the organisms and their general physical and nutritional needs.

The panel wishes to thank the many scientists who contributed information. Special appreciation is expressed to Marcia A. Duncan, research assistant; Mary Jane Engquist, staff assistant; and to Dorothy M. Woodbury and Cicely Henry, who served as administrative secretaries, for preparing draft documents for the meeting and for producing the final manuscript of this report; they have been most helpful. The panel also acknowledges the help of Diosdada DeLeva, Maryalice Risdon, and Wendy D. White for bibliographic editing and Harry Hatt, of the American Type Culture Collection, for standardizing the nomenclature of microorganisms.

The final report was edited and prepared for publication by F. R. Ruskin, for whose assistance the panel is grateful.

Contents

INTRODUCTION 1

1 RAW MATERIALS FOR MICROBIAL PROCESSES 10
Typical Raw Materials 11
Underutilized Raw Materials 15

2 FOOD AND ANIMAL FEED 18
Food Preservation 19
Improving Nutritional Value 24
Production of Meat-Like Flavors 28
Koji Method of Producing Enzymes 30
Indonesian Tempeh 32
Single-Cell Protein Production 37

3 SOIL MICROBES IN PLANT HEALTH AND NUTRITION 47
Mineral Cycling by Soil Microorganisms 48
Mycorrhizal Fungi 51
Biological Control of Soil-Borne Pathogens 55

4 NITROGEN FIXATION 59
Symbiotic Systems 61
Asymbiotic Fixation 71

5 MICROBIAL INSECT CONTROL AGENTS 80
Development of Bioinsecticides 80
Bacteria 84
Viruses 89
Protozoa 94
Fungi 98

6 FUEL AND ENERGY 107
Ethanol 108
Utilization of Cellulose 111
Methane 111

Methanol 116
Hydrogen 117
Bacterial Leaching 119

7 WASTE TREATMENT AND UTILIZATION 124
Algal-Bacterial Systems 125
Composting 131
Anaerobic Lagoons 133
Recycling Animal Waste by Aerobic Fermentation 136
Recycling Animal Waste by Anaerobic Fermentation 138

8 CELLULOSE CONVERSION 142
Volvariella Species 143
Lentinus edodes 144
Pleurotus Species 148
Thermoactinomyces Species 149
Phanerochaete chrysosporium 151
Trichoderma reesei 152
Other Species 153

9 ANTIBIOTICS AND VACCINES 158
Antibiotics 160
Vaccines 169

10 PURE CULTURES FOR MICROBIAL PROCESSES 177
Major Pure Culture Collections 178
World Data Center and Microbiological Resource Centers 178
Preservation Methods 180
Mixed Microbial Cultures 184
Patenting of Processes Involving Microorganisms 184

11 FUTURE PERSPECTIVES IN MICROBIOLOGY 186

REGULATIONS FOR PACKAGING AND SHIPPING VIABLE MICROBIAL CULTURES 191

Board on Science and Technology for International Development 193
Advisory Committee on Technology Innovation 195

Introduction

Microorganisms have simultaneously served and assaulted man throughout history. Man is totally dependent on some microbes for life processes, while remaining subject to the destructive capacities of others in diseases not yet conquered.

The study of microorganisms and microbial processes has provided a variety of benefits. For instance:

- World health has been improved through the discovery of the microbial causes of most human, animal, and plant diseases, leading to the development of vaccines, antibiotics, and chemical agents to combat many of these diseases.
- Foods have been improved in quality and protected from spoilage to enable wide distribution and storage against times of need.
- Sewage treatment methods have been developed to break the chain of disease transfer through waterborne pathogens. Microorganisms also enhance the water quality of rivers and lakes by degrading naturally occurring organic matter.
- Farming practices have been improved through recognizing and capitalizing on the role of soil microorganisms; microbes have been used to break down nonedible crop residues for reuse by new crops. Nitrogen-fixing microorganisms have been used to inoculate legumes.
- Microbial fermentation processes have provided foods, beverages, medicines, and chemicals for human use.

Microbes, as organized systems of enzymes, can often perform these functions more efficiently than purely chemical processes, and current environmental and economic constraints make the potential contribution of microbes increasingly attractive.

From these examples it is clear that microbes can be marshaled to aid in solving many important global problems including food shortages, resource recovery and reuse, energy shortages, and pollution. Microbiology is particularly suited to make important contributions to human needs in developing countries, yet it has received comparatively little attention. The range of possible applications covers uses by individuals and industries in rural settings, villages, and cities.

This report covers examples of microbial processes that may be useful in developing countries. Although many of these processes may not have a direct and immediate use, their scope and diversity should serve to indicate the strong potential for microbial applications.

Above all, the report highlights the pervasiveness and importance of microbes, along with the increasing need to train microbiologists and to support their research and development activities. A group of well-trained microbiologists with adequate support can make valuable contributions to social welfare.

Organisms Involved in Microbial Processes

The organisms responsible for the microbial processes discussed in this report are an integral, all-pervasive part of the biological world. Although they are rarely seen (the larger fungi, mushrooms, are perhaps the most visible), it is estimated that microorganisms make up about one-quarter of the biomass—the total weight of living organisms in the world—with animals and plants accounting for the remainder. Microorganisms occur everywhere, and extraordinary aseptic measures are required to exclude them from places where their presence would be harmful, such as the operating room of a hospital or a food-processing plant. Even then, these measures are not always successful.

The bulk of microorganisms reside in the soil, where they are responsible for the predominant biological activity. Others are located in the upper layers of the oceans and in fresh and brackish waters, as well as on the surfaces above ground, in the air, and of course inside larger organisms, both plant and animal.

A number of microorganisms are harmful, or pathogenic, to humans and animals. Although the terms microbe or germ initially were used to describe any minute microorganism, they tend to be used especially to connote harmful organisms. Yet most microorganisms are either harmless or essential for the maintenance of the biological cycles on which all life depends.

Microorganisms comprise the following classes of organisms:

- Bacteria
- Fungi (yeasts and molds)
- Algae
- Protozoa
- Viruses.

Their classification, characteristics, and harmful and beneficial effects are shown in Table 1.

Physicochemical Factors Affecting Microbial Growth

A number of physical factors affect the growth or retardation of microorganisms, including temperature, osmotic pressure, acidity or alkalinity, the presence of oxygen or light, and the degree of agitation. Although no species of microbes can survive over the complete range of conditions found in nature, there are varieties that thrive in hot springs, polar wastes, acidic bogs, and highly saline waters like the Dead Sea.

Temperature

Most microorganisms grow within a temperature range of 30°C. Individual species have well-defined upper and lower temperature limits and optimum temperatures for growth.

Microorganisms are usually divided into three groups with respect to their most favored temperature range. Psychrophiles grow best between about 0°C and 30°C. These organisms occur in cold areas and are frequently associated with refrigerated food spoilage. Mesophiles grow best between about 20°C and 50°C. Most disease-causing bacteria are in this group. Thermophiles grow best from 40°C to 70°C. This division into three groups is convenient but somewhat arbitrary, since the dividing lines are not sharp. Further, not every organism can grow over the entire range indicated for its group.

Acidity and Alkalinity

Taken as a whole, microorganism species can tolerate a wide range of acidity and alkalinity. Some thrive under highly acidic conditions (pH 1–3) and others in alkaline environments (pH 9–10). However, most microorganisms grow best at neutral pH (pH 7).

Oxygen

Microorganisms can be divided into three major groups with respect to their oxygen requirements. Obligate aerobes have a requirement for oxygen and grow best when oxygen is continuously available. Obligate anaerobes grow in the absence of free oxygen. The requirement for oxygen reflects the metabolic pathways the organisms use to obtain energy. Aerobes break down

TABLE 1 Microorganisms: Characteristics, Problems, and Uses

Organisms	Characteristics	Problems	Uses
Bacteria: Schizomycetes or Protista	Single-celled; spherical rod and spiral forms. Most are saprophytes (use dead matter for food).	Some forms are pathogens for plants, humans, and animals.	Break down organic matter and assist soil fertility, waste disposal, and biogas production; source of antibiotics and other chemicals.
Fungi: Plants devoid of chlorophyll	Variety of forms; microscopic molds, mildews, rusts, and smuts; larger mushrooms and puffballs.	Rot textiles, leather, harvested foods, and other products; cause important plant and animal diseases.	Assist in recycling cellulose, lignin, and other complex plant constituents; mushrooms and yeasts are important in food and nutrition; many are also used in chemical and pharmaceutical industries.
Algae: Thallophyta (undifferentiated plants)	Single cells, colonies, or filaments containing chlorophyll and other characteristic pigments. No true roots, stems, or leaves; aquatic.	Cover pond surfaces, producing scum and unpleasant odor and taste (in drinking water); absorb O_2 from ponds and some produce toxins.	Red and brown seaweeds are important foods in Asia and Polynesia. Red algae produce agar. Some blue-green algae fix nitrogen. Major food for ocean fish.
Protozoa: Microscopic animals	Single-celled or groups of similar cells, found in fresh and sea water, in soil, and as parasites in animals, man, and some plants.	Responsible for serious human and animal diseases–malaria, sleeping sickness, dysentery, etc.	Assist in breakdown of organic matter such as cellulose in ruminant nutrition.
Viruses: Submicroscopic forms; considered intermediate between living and nonliving.	Infective agents, capable of multiplying only in living cells; composed of proteins and nucleic acids.	Cause of variety of diseases in humans (measles, influenza, pneumonia, poliomyelitis), animals (foot and mouth, canine distemper), and in plants.	Important as carriers of genetic information. Also cause diseases in insects and other pests, and research is directed to their use in biological pest control.

their nutrients by a sequence of enzyme reactions that require oxygen. Anaerobes utilize a pathway for metabolism that does not require free oxygen, and in fact they may be inhibited by it.

The third group of organisms are the facultative anaerobes. These can use either metabolic pathway, depending on the presence or absence of oxygen.

Osmotic Pressure

The osmotic pressure across a cell wall depends on the relative concentration of dissolved substances within the cell and outside it. For example, most bacteria can grow over a broad range of salinity because their cells are capable of maintaining a relatively constant internal salt concentration. But if salt concentrations outside the cell become too high, water is lost from the cell and growth is inhibited. This is the basis for food preservation by salt. Sugars and other substances also influence osmotic relationships between cells and their environment.

Nutritional Requirements for Microbial Growth

All microorganisms require water to grow and water can be considered the single most important component in their growth.

Microorganisms can be divided into two groups based on the source of carbon they convert into their cell components. Heterotrophic organisms utilize organic compounds as a source of carbon for both synthesis and energy. Autotrophic organisms utilize carbon dioxide as their major source of carbon for synthesis and obtain energy either from the sun (through photosynthesis) or by metabolizing inorganic compounds. The inorganic compounds that can be used by various autotrophic organisms include ammonia, hydrogen, reduced iron, manganese and other minerals, and hydrogen sulfide.

Heterotrophs can utilize a wide variety of organic materials as sources of carbon. In fact, there are probably no biologically generated materials in the environment that cannot be degraded by some species of microorganism.

In addition to carbon, all organisms require sources of the other elements found in cell components. These include nitrogen, sulfur, phosphorus, and potassium. Both heterotrophs and autotrophs require certain inorganic salts for optimum growth and reproduction.

Most microorganisms cannot utilize (fix) atmospheric nitrogen and require nitrogen in the form of an ammonium or nitrate salt or in an organic form. Sulfur is usually obtained from sulfate salt and phosphorus from salt of phosphoric acid.

Raw Materials for Microbial Processes

A variety of materials have been used in microbial processes in industrialized countries. For less-developed nations, however, it is not necessary to restrict usage to these substances; indigenous raw materials, for example agricultural residues, may be much more appropriate.

Food and Animal Feed

Microorganisms have long been used to produce certain foods, beverages, condiments, and animal feeds. Recently, several new commercial microbial processes have been developed. These include the production of single-cell protein to supplement animal feeds; mushrooms for human food from agricultural wastes; microbial rennet for cheese making; enzymes such as glucose isomerase; meat-like flavorings using the Chinese soy sauce and Japanese miso processes; xanthan and amino-, hydroxy-, and keto-acids, and vitamins, among other products.

There are many potential ways for utilizing microorganisms in food production, from the household and village level to full-scale commercial operations. The need continues for better food preservation and methods to reduce postharvest food spoilage.

Soil Microbes in Plant Health and Nutrition

The region where the roots of plants make contact with the soil is called the rhizosphere. This is a complex biological area in which the microbial population is considerably higher and its activity greater than in root-free soil. Growth of microorganisms in the rhizosphere is undoubtedly enhanced by nutritional substances released from the roots, and growth of plants is influenced by microbial metabolic products released into the soil.

Of great significance are certain fungi that infect roots and form mycorrhizae. These fungi can absorb and translocate phosphate and other essential nutrients and make them available to plants. With a greater need for food for an ever-growing population, increased attention should be given to the effects of the rhizosphere on plant nutrition.

Nitrogen Fixation

As demands for fertilizer increase, and as the energy crisis becomes more acute, greater attention must be given to microbial fixation of atmospheric nitrogen. The emphasis should be on applying known technology, of which legume inoculation to increase crop production is a good example. Basic research on culture and ecology of both symbiotic and nonsymbiotic nitrogen-fixing microorganisms could lead to an increase in the world's supply of edible protein. This would be of even greater significance if microorganisms that fix nitrogen, or their nitrogen-fixing genes, could be transferred to microorganisms that can be established in nonleguminous crops, such as rice and other cereals, so they could utilize nitrogen from the air. Cultivation of free-living nitrogen-fixing blue-green algae that grow in nitrogen-deficient substrates is another goal. The potential for development in these areas is great.

Microbial Insect Control Agents

In the search for safe, alternative methods of controlling insect pests, the use of microorganisms that cause disease in insects offers distinct possibilities. Insects, like humans, animals, and plants, are susceptible to microbial diseases. Microbes that produce diseases in insects are termed entomopathogens. In many cases they can significantly reduce natural populations of insects. Safety, specificity, effectiveness, and cost are the decisive considerations in the development of any insecticide. A number of entomopathogens fulfill these criteria and are therefore potentially useful bioinsecticides. Some are already being produced commercially, and more are in development.

Fuel and Energy

Most nations today are facing shortages of fuel and energy. Yet if development is to proceed, increasing amounts of energy will be required. To meet these growing requirements, attention must be directed to the development of unconventional and renewable energy systems.

Microbial processes already help provide energy. In the countries of South and Southeast Asia and in the People's Republic of China, for example, many small farms and villages are using methane generators that utilize fermented animal manure, human wastes, and other waste substances to produce "biogas" for household cooking, lighting, and power. In some countries alcohol produced by microbial fermentation is added to petroleum products to supplement scarce fuel supplies. These processes that depend upon the solar-produced biomass may hold unique promise for supplying some of the energy requirements of less-developed nations. The microbiological conversion of plant matter into fuel circumvents the millions of years required for plant material to become fossil fuel through natural processes.

Waste Treatment and Utilization

A number of water and wastewater purification processes utilize microbes. Many opportunities exist for waste utilization and recycling, including refeeding of animal wastes; algal farming for fish culture and as a source of animal feed and fermentable substrates; and the upgrading of cellulose wastes by protein enrichment for use as fodder.

Cellulose Conversion

Cellulose, a renewable resource from agricultural and forestry products, is a major component of many solid wastes and residues. Usually, cellulose is bound to lignin. The lignocellulose complex is a substrate that must be chemically degraded before the cellulose can be used in some commercial processes.

Cellulose can be degraded by chemical or enzymatic hydrolysis to soluble sugars. These sugars can then be used by microbes to form ethanol, butanol, acetone, single-cell protein, methane, or other products of fermentation. In some cases, cellulose can be converted directly into these products by fermentation. The technology for refined cellulose degradation is readily available for recycling paper, cardboard, etc.

Biomass agriculture and forestry may hold great economic potential for certain less-developed countries, particularly in tropical and subtropical regions.

Antibiotics and Vaccines

Although approximately 4,000 antibiotics are known, most have no practical value because of their toxicity to human beings, lack of efficacy, or high production cost. There are only about 50 widely used antibiotics. Extensive use of antibiotics in medicine began in 1945 with penicillin. Currently, antibiotics are widely used in human and veterinary medicine, and to a lesser extent in agriculture, where they are used to increase the weight of livestock and poultry, to control plant diseases, and as insecticides. New antibiotics are being sought and old ones are being modified to improve their properties.

Killed, attenuated, or living microorganisms, or their products, have been used for many years to produce immunity against certain human diseases such as smallpox, cholera, yellow fever, tetanus, and diphtheria. Additional research is needed to improve these vaccines and to produce new ones. Special emphasis should be placed on effective programs and delivery systems for existing vaccines.

Pure Cultures for Microbial Processes

Microorganisms are an extremely important natural resource. Because of the present and potential usefulness of beneficial microorganisms, it is essential that their germ plasm be preserved, just as plant germ plasm is preserved in seed banks and endangered animal life is protected in various ways.

Several outstanding culture collections of microorganisms exist today, and they are essential to research and teaching in microbiology as well as commercial microbial production.

Chapter 1

Raw Materials for Microbial Processes

Microorganisms, like all other forms of life, require water and nutrients for growth, reproduction, and maintenance. In addition to suitable sources of *utilizable* carbon, nitrogen, and sulfur, microbes generally require sodium, potassium, phosphorus, iron, and other minerals. The major factor in selecting raw materials for microbial processes is the source of carbon.

Microbial processes have long been harnessed for the benefit of man in the production of foods, medicines, and alcoholic beverages. Nature employs microbes on a much grander scale to establish and maintain a balance among the diverse forms of life on this planet. The underlying agents responsible for the myriad syntheses, transformations, and other reactions caused by microbes are the enzymes—biological catalysts of high specificity and efficiency.

One important aim of science and technology has been to domesticate beneficial microbes, especially for the transformation of raw materials to worthwhile end products.

In general, most raw materials are naturally occurring substances from which more useful materials can be produced. In this sense, microbes themselves may be considered raw materials suitable for further processing. The use of microbes as single-cell protein (SCP) is an example (see Chapter 2). In this report the discussion will be limited to major carbon sources found in nature, formed mostly by plants through photosynthesis, which can be used either for producing additional biomass (e.g., SCP) or for further transformations (e.g., alcohol).

In theory, any abundant carbon source might be employed for microbial processes, including coal, petroleum, lignocellulose, starch, sugar, organic acids, and even carbon dioxide. Some of these sources are currently used; others, such as coal and carbon dioxide, present considerable technological barriers. Coal would have to be converted first to a readily usable carbon base (perhaps paraffin or methanol) because it is biologically inert and may contain compounds potentially inhibitory or toxic to microbes. However, plant biomass and, to a lesser extent, animal biomass, represent utilizable sources of carbon for microbial processes. Well-known examples of microbial processes based on these sources are the production of alcohol from grain and cheese from milk.

Although carbon dioxide is a form of carbon that can be assimilated by some microorganisms, this raw material is utilized mainly by plants through the mechanism of photosynthesis. Primary photosynthetic productivity (growth of plants using solar energy) of the earth has been estimated to be 155×10^9 t* of material per year on a dry weight basis.

The distribution of plant biomass produced by photosynthesis is shown in Table 1.1. Land-based plants account for 65 percent of the weight of the biomass produced annually, even though they occupy only about 29 percent of the area. The dominant yearly production of land-based biomass, approximately 42 percent, is produced as forest.

Although agricultural crops account for only 6 percent of the primary photosynthetic productivity, they provide not only a vital portion of food for man and animals, but other essentials such as structural materials, textiles, and paper products as well. Agricultural raw materials are the most important source of carbon for microbial conversion processes. The historical multipurpose use of agricultural crops has maintained a continued heavy dependence on this source.

Most agricultural crops and residues are relatively free from toxic materials and this, in addition to their availability, may have stimulated their use as a raw material for microbial processes. Because of these advantages, along with the real technological barriers to using other carbon sources, agricultural crops and residues can be expected to retain their dominance as the carbon source for microbial processes. But this must not preclude the further exploration and exploitation of other sources, especially those indigenous to the less-developed nations. In addition, based on their unique environments and requirements, certain countries may have an opportunity to establish new plants and practices for microbial processes.

Typical Raw Materials

Some typical raw materials and fermentation products used in developed countries are listed in Table 1.2. Selected combinations of other materials are used as substrates for various products. These raw materials provide carbon, nitrogen, salts, trace elements, vitamins, and other requirements for the processes; they are few in number because the conditions for large-scale microbial processes impose limitations on the materials that may serve as substrates.

In general, the raw materials mentioned in Table 1.2 are traditionally used for microbial processes because of their suitability for specific processes. But

*In this report t represents metric ton.

in another geographical area, or for new or different processes, one need not be limited to what has been used in the past. Table 1.3 lists a variety of important food crops grown in developing countries. These crops or their residues may also be considered as raw materials for microbial processes.

A variety of other common waste materials derived from agricultural, forest, and urban sources, may serve as substrates for microbial processes (Table 1.4).

TABLE 1.1 Estimated Primary Photosynthetic Productivity of the Earth

Area (total = 510 million km^2)				Net Productivity (total = 155.2 billion tons dry wt/yr)		
Total Earth	100%			100%		
Continents	29.2			64.6		
Forests		9.8			41.6	
Tropical Rain			3.3			21.9
Raingreen			1.5			7.3
Summer Green			1.4			4.5
Chaparral			0.3			0.7
Warm Temperate Mixed			1.0			3.2
Boreal (Northern)			2.4			3.9
Woodland		1.4			2.7	
Dwarf and Scrub		5.1			1.5	
Tundra			1.6			0.7
Desert Scrub			3.5			0.8
Grasslands		4.7			9.7	
Tropical			2.9			6.8
Temperate			1.8			2.9
Desert (Extreme)		4.7			0	
Dry			1.7			0
Ice			3.0			0
Cultivated Land		2.7			5.9	
Freshwater		0.8			3.2	
Swamp & Marsh			0.4			2.6
Lake & Stream			0.4			0.6
Oceans	70.8			35.4		
Reefs & Estuaries		0.4			2.6	
Continental Shelf		5.1			6.0	
Open Ocean		65.1			26.7	
Upwelling Zones		0.08			0.1	

Source: James A. Bassham. 1975. Cellulose as a chemical and energy resource. In *Cellulose as a chemical and energy resource.* New York: John Wiley and Sons.

TABLE 1.2 Typical Raw Materials and Products in Industrialized Countries

Raw Materials	Products
Sulfite Waste Liquor	Single-Cell Protein (SCP)
Ethanol	SCP Acetic Acid
Methanol	SCP
Whey	SCP Lactic Acid
Paraffins	SCP Citric Acid Amino Acids
Molasses	Ethanol Glutamic Acid

TABLE 1.3 Estimated Production of Major Food Crops in Developing Countries*

Crop	Metric Tons (in thousands)	Percent
Paddy	186230	21.36
Cassava	103486	11.87
Wheat	95048	10.90
Maize	73328	8.41
Banana/Plaintain	55199	6.33
Coconuts	32664	3.75
Sorghum	31173	3.57
Yams, Taro, etc.	28777	3.30
Potatoes	26909	3.09
(Pulses)**	25997	(2.98)
Citrus	22040	2.53
Millet	21452	2.46
Barley	20775	2.38
Sweet Potatoes	17630	2.02
Soybeans	13842	1.59
Groundnuts	13502	1.55
Tomatoes	12755	1.46
Grapes	12720	1.46
Mangoes	12556	1.44
Watermelon	10436	1.20
Dry Beans	8537	0.98
Onions	6474	0.74
Percentage of Total Developing Country Food Crop Production		94.39

*Developing market economies as defined in the *FAO Production Yearbook* (1977).

**Pulses–total legumes except soybeans and groundnuts.

Source: *FAO Production Yearbook*. New York: UNIPUB, 1977.

TABLE 1.4 Typical By-product Substrates for Use in Microbial Processes in Developing Countries

Agricultural	Other
Molasses	Animal Manures
Maize Stover	Sewage
Straw	Municipal Garbage
Bran	Paper Mill Effluent
Coffee Hulls	Cannery Effluent
Cocoa Hulls	Fishery Effluent
Coconut Hulls	Slaughterhouse Effluent
Fruit Peels	Milk-Processing Effluent
Fruit Leaves	
Bagasse	
Oilseed Cakes	
Cotton Wastes	
Tea Wastes	
Bark	
Sawdust	

Of the lists of materials in Tables 1.2, 1.3, and 1.4, a few may hold special promise for use in the microbial processes that occupy the bulk of this report. The criteria for selecting these raw materials for research are their almost year-round availability in large volume in many developing areas and their ease of assimilation by microorganisms.

Among raw materials commonly used for microbial processes (Table 1.2), molasses is probably the one most readily available for use as a substrate in developing countries. Because it contains both easily assimilable sugars and necessary micronutrients, it is a very useful substrate that is readily utilized by a variety of microorganisms. However, it contains very little nitrogen. Starch (or syrups produced from starch) is also a good substrate, and many potential sources are available. These include the cereal crops (maize, rice, wheat, etc.) and starchy tubers such as potato and cassava.

In addition to the crops that may be good sources of starch, a few of the plentiful products and waste products listed in Tables 1.3 and 1.4 may merit particular consideration for some of the microbial processes. Cassava, for instance, may be a good choice as a substrate to produce ethanol, SCP, and other economically valuable substances. This might be a better disposition of the crop than its present widespread use for food, since its low protein-to-calorie ratio makes it less than ideal nutritionally. Another promising raw material is coffee-processing waste, which is produced in large amounts (4.5 t of by-product for each t of dehulled coffee). It appears to be a good substrate for the growth of various fungi and yeasts. Taro (*Colocasia esculenta*), though a well-known staple food, has a more limited distribution than some of the agricultural products mentioned above, existing as a commercial crop only in Egypt, West Africa, Southeast Asia, and some Pacific and Caribbean Islands.

However, it, too, has potential as a substrate.

Domestic sewage or industrial wastes offer possible fermentation substrates for algae or bacteria. Unless the supply is properly planned, however, it cannot always be counted on to meet the demands of a large microbial process.

The single most abundant potential source of carbon—in developing countries and elsewhere—is cellulose (see Chapter 8). This is a constituent of many foods, fiber crops, agricultural residues, and wood and forest residues, some of which are mentioned in Table 1.4.

Most of these selected substrates are rich in carbohydrates, but for certain microbes they may have to be supplemented with sources of nitrogen, salts, trace metals, and other requirements. Possible sources for some of these supplements in developing countries may be whole yeast or distiller's dried solubles from local alcoholic fermentations. In some countries meat or fish by-products such as slaughterhouse wastes or gutting and canning residues would be excellent supplements. Combinations of plants might also be used to meet the nutritional requirements of producing organisms. For example, cassava (carbohydrate) could be combined with soybeans (high nitrogen).

Underutilized Raw Materials

The materials so far discussed as possible substrates for microbial processes are widely cultivated and available. But there are many less-known plants, or plants that may be used only locally, that may be excellent candidates for this purpose. An example is the winged bean (*Psophocarpus tetragonolobus*), now becoming more popular as a food in Southeast Asia and West Africa because of its unique combination of protein-rich and edible seeds, tubers, and leaves.

Certain tropical plants, such as basella and amaranths, have not received much attention as food sources, but they may give a greater yield than many crops in extensive use and may also be useful as substrates. However, underdeveloped raw materials selected for large-scale microbial processes will probably have to meet the requirements discussed in the Introduction.

Plants can also be grown specifically for biomass as a fermentation substrate. Plants selected for such use should grow and reproduce rapidly, contain a low crystallinity cellulose and a low lignin content, and be easily harvested and transported. Another desirable characteristic of plants for biomass would be an ability to grow in ecological niches in which they will not compete with or eclipse regular crops.

For instance, the buffalo gourd (*Cucurbita foetidissima*) does well in arid conditions, and the salt bushes (*Atriplex* spp.) and tamarugo (*Prosopis tamarugo*) are salt tolerant and might be introduced into countries with arid and saline areas. Aquatic plants such as the reed (*Phragmites communis*), cattails (*Typha* spp.), the papyrus reeds (*Cyperus* spp.), mat rush (*Juncus effusus*), textile screw pine (*Pandanus tectorus*), and eel grass (*Zostera marina*) are examples of plants that grow in a saline aquatic environment.

The use of the water hyacinth (*Eichornia crassipes*) to treat domestic sewage with large quantities of plant biomass produced during the process is receiving increased attention because of the tremendous growth rate of this plant—as much as 850 kg/ha/day of dry plant material has been reported. This prolific growth has made the water hyacinth a troublesome weed in certain tropical and semitropical areas, particularly the Nile Basin and southern United States. Biomass from domestic waste treatment can be used to produce biogas and fertilizer, feed, or a protein concentrate. At present, domestic wastewater treatment using the water hyacinth is being demonstrated in the United States.

Kelps and seaweeds are sources of carbohydrates other than cellulose. These plants have the drawback of high water and salt content. Farming such plants and harvesting them economically may also present special problems.

Countries with a limited supply of oil and natural gas are not likely to consider petroleum hydrocarbons as a microbial substrate. But countries with large oil and gas deposits also have a supply of methane, which may be used as a carbon and hydrogen substrate for the growth of organisms for single-cell protein.

Research Needs

In many developing countries there is need to establish basic information about potential substrates for microbial processes by:

- Systematic identification of resources;
- Analysis of constituents and properties of promising individual sources;
- Identification of scientific, technological, and institutional resources and constraints; and
- Determination of optimal process based on best use of substrate, economic justification, available technical support, and unique area needs.

References and Suggested Reading

Bassham, J. A. 1975. Cellulose as a chemical and energy resource. In *Cellulose as a chemical and energy resource: Cellulose Conference Proceedings*, held under the auspices of the National Science Foundation, at the University of California, Berkeley, June 25-27, 1974, C. R. Wilke, ed., pp. 9-19. New York: John Wiley and Sons.

Birch, G. G.; Parker, K. J.; and Worgan, J. T., eds. 1976. *Food from waste.* London: Applied Science Publishers Ltd.

Conference on Capturing the Sun through Bioconversion, Proceedings, sponsored by the United States Energy Research Development Administration and others, Washington, D.C., March 10-12, 1976. Washington, D.C.: Washington Center of Metropolitan Studies.

Food and agriculture: readings from Scientific American. 1976. San Francisco: W. H. Freeman and Company.

National Academy of Sciences. 1975. *Underexploited tropical plants with promising economic value.* Report of an *Ad Hoc* Panel of the Advisory Committee on Technology Innovation, Board on Science and Technology for International Development, Commission on International Relations. Washington, D.C.: National Academy of Sciences.

______. 1976. *Making aquatic weeds useful: some perspectives for developing countries.* Report of an *Ad Hoc* Panel of the Advisory Committee on Technology Innovation, Board on Science and Technology for International Development, Commission on International Relations. Washington, D.C.: National Academy of Sciences.

______. 1976. *Renewable resources for industrial materials.* Report of the Committee on Renewable Resources for Industrial Materials, Board on Agriculture and Renewable Resources, Commission on Natural Resources. Washington, D.C.: National Academy of Sciences.

______. 1977. *Methane generation from human, animal, and agricultural wastes.* Report of an *Ad Hoc* Panel of the Advisory Committee on Technology Innovation, Board on Science and Technology for International Development, Commission on International Relations. Washington, D.C.: National Academy of Sciences.

Perlman, D. 1977. Fermentation industries, *quo vadis*? *Chemical Technology* 7:434-443.

Schlegel, H. G., and Barnea, J., eds. 1977. *Mircrobial energy conversion*. Oxford: Pergamon Press.

White, J. W., and McGrew, W., eds. 1977. *Clean fuels from biomass and wastes.* Proceedings of the symposium held on January 25-28, 1977, at Orlando, Florida, sponsored by the Institute of Gas Technology. Chicago: Institute of Gas Technology.

Wilke, C. R., ed. 1975. *Cellulose as a chemical and energy resource*: *Cellulose Conference Proceedings*, held under the auspices of the National Science Foundation, at the University of California, Berkeley, June 25-27, 1974. New York: John Wiley and Sons.

Research Contacts

Carl-Göran Hedén, Karolinska Institutet, Solnavagen 1, S-10401 Stockholm 60, Sweden.

Arthur E. Humphrey, School of Engineering, University of Pennsylvania, Philadelphia, Pennsylvania 19104, U.S.A.

F. K. E. Imrie, Tate and Lyle Ltd., Philip Lyle Memorial Research Laboratory, University of Reading, P. O. Box 68, Reading RG6 2BX, England.

David Perlman, School of Pharmacy, University of Wisconsin, Madison, Wisconsin 53706, U.S.A.

Steven R. Tannenbaum, Department of Nutrition and Food Science, Massachusetts Institute of Technology, Cambridge, Massachusetts 02139, U.S.A.

Noel D. Vietmeyer, National Academy of Sciences, 2101 Constitution Avenue, N.W., Washington, D.C. 20418, U.S.A.

Daniel Wang, Department of Nutrition and Food Science, Massachusetts Institute of Technology, Cambridge, Massachusetts 02139, U.S.A.

J. T. Worgan, National College of Food Technology, University of Reading, St. George's Avenue, Weybridge, Surrey KT13 0DE, England.

Chapter 2

Food and Animal Feed

At least 25 percent of the world's population, approximately one billion people, suffer from hunger and malnutrition. It is vital, therefore, that they produce more food and improve their standard of nutrition. Food losses should be decreased through low-cost methods of food preservation, since standard modern methods of food processing such as canning, freezing, or dehydration with artificial heat make the preserved food too expensive for families on an income of US$200–$300 per year.*

Most cultures have traditionally used some form of microbial process to preserve foods that would otherwise spoil. Some of these processes have also contributed to increasing the nutritive value of the final product through the increased production of essential nutrients or the synthesis of nutrients not present in the original food. Cheeses are perhaps the most widespread and best known of these foods. But there are many others, some of which are largely confined to certain parts of the world and almost totally unknown elsewhere. This section describes a number of these processes.

While it is recognized that food habits and customs are among the most difficult to change, there have already been dramatic changes in some food habits of developing countries. The spread of wheat flour bread throughout much of the lowland tropics is an example of one such change, and it involves a microbial process, yeast fermentation.

The purpose of this section is to bring to the attention of the reader a number of lesser-known processes that preserve or enhance the nutritive value of foods and beverages and that may merit introduction to areas where they are unknown or invite more widespread use and improvement where they are already used. None of the methods (except for large-scale single-cell protein production of animal feeds) requires large investment in capital equipment or plants.

These processes also illustrate the potential contribution of microbiology toward diversifying the use of limited resources as well as the need to develop

*National Academy of Sciences. 1978. *Postharvest Food Losses in Developing Countries*. Washington, D.C.

trained microbiologists for a variety of research and development opportunities. It must be stressed that food processing is always potentially hazardous, and new methods or products should only be marketed after careful investigation by qualified personnel.

Food Preservation

Three relatively low-cost methods of food preservation frequently carried out in combination are: 1) salting; 2) sun or smoke drying; and 3) acid fermentation.

Salt is one of the best low-cost chemicals for preserving fresh foods ranging from vegetables to meats and fish. Added in concentrations of approximately 2–2.5 percent by weight to fresh vegetables, salt promotes an anaerobic lactic acid fermentation. The fermentation favors development of a microbial flora that converts, for example, cucumbers to pickles and cabbage to sauerkraut (see flow sheets). Pickling is the traditional method of providing a winter supply of vegetables in much of Asia and Europe. There is less need for extended storage in tropical areas where vegetables can be grown on a year-round basis, but these methods may be valuable for taking advantage of periodic surplus production and providing variety to the diet.

Korean kimchi, a staple in that country, consists of mixed vegetables fermented by lactic acid bacteria. An inoculum is generally not required, as the bacteria are ubiquitous. The salt concentration, coupled with anaerobic conditions at ambient temperature, controls the development and sequence of the lactic-acid-producing organisms. The combination of salt and acid leads to products with excellent keeping quality.

Fish Sauce and Paste

Fish and shrimp are excellent protein foods, but they are highly perishable. Also, many fish caught in nets are too small to be sold commercially or are species not generally consumed directly as fresh fish. Thus, low-cost methods of preserving small and surplus fresh fish and shrimp are of great importance in food distribution and consumption. Larger surplus fresh fish are often salted, sun dried, and consumed as dried fish or as fish powder.

Large quantities of small fish are fermented to produce fish or shrimp sauces and pastes in Southeast Asia. The basic procedure is to mix the freshly netted small and trash fish with sea salt, in proportions that ensure that the extracted fish juices contain about 20 percent salt in the final product. Such high salt concentrations inhibit putrefaction. No inoculum is required: the microorganisms in the gut of the fish and enzymes in the fish tissues control hydrolysis (solubilization) of the fish proteins.

FLOW SHEET: Cucumber (Pickle) Fermentation

↓

Fresh Cucumbers

↓

Wash

↓

Cover with 5% salt brine (add dill or other flavoring)

↓

Cover fermentation container to avoid evaporation and to exclude air

↓

Ferment for 1 or 2 weeks
(total acid 0.6–1.0% as lactic acid, pH 3.6–3.4)

FLOW SHEET: Cabbage (Sauerkraut) Fermentation

Fresh Cabbage*

↓

Trim and clean

↓

Remove core

↓

Shred (to 2–5 mm width)

↓

Add salt (about 2.25% by weight) and distribute evenly

↓

Fill into fermentation containers
Salt and shredded cabbage can be mixed as fermentation container is filled.

↓

Cover top of fermentation container and use a water seal to prevent entrance of air.
The seal must allow escape of CO_2 gas produced during fermentation.

↓

Ferment to acidity desired
For optimum keeping quality the acidity should be high (below pH 4), but the sauerkraut can be consumed earlier if desired.

The fish sauces (nuŏc mắm in Vietnam, patis in Indonesia and the Philippines, and nampla in Thailand) are salty condiments adding some essential amino acids and vitamins (mainly B-complex) to the diet. Similar processes in which less protein hydrolysis occurs lead to fish or shrimp pastes. The pastes may be mixed with cereals; ragi and millet are reported to be used for this purpose. These products are of limited nutritional value (particularly for infants)—despite their protein content—because of their high salt content.

*Or other green leafy vegetables.

Acid Milks, Yogurt, and Cereal

Fresh unpasteurized milk, if allowed to stand at ordinary temperature, sours naturally because the streptococci and lactobacilli present convert milk lactose to lactic acid. This results in a natural acid preservation of the nutrients. The acid produced results in stable products resistant to putrefaction and the development of food spoilage or some disease-producing organisms. It is, however, susceptible to some fungi and especially to *Geotrichum*.

Modern yogurt processing involves inoculating milk with *Streptococcus thermophilus* and *Lactobacillus bulgaricus* and incubating it at 45°C. For the production of Russian kefir, milk inoculated with kefir grains (consisting of a lactobacillus and a yeast growing symbiotically) yield an acidified, carbonated sour milk with a low alcohol content. Incubation is at room temperature.

Sour milks or yogurts boiled with ground whole wheat or bulgar wheat and then sun dried yield extremely nutritious stable foods, which can be stored for years without deterioration. This is the basic process for Egyptian kishk and Greek trahana. Other ingredients such as spices, pepper, tomato, onion, or garlic, and other vegetables may also be incorporated in the products, which are either consumed directly or used as a major protein ingredient in soups.

Indian Idli (Dosai)

Indian idli is a nutritious, protein-rich acidic steamed bread popular in South India. Its acidity makes it quite resistant to food spoilage and certain disease-producing organisms.

In preparing idli, polished rice and dehulled black gram (*Phaseolus mungo*, mung bean, a legume similar to split pea) are soaked separately during the day. The proportions can be any combination from 1 to 3 parts rice to 1 part black gram. Since black gram is more expensive than rice, most poorer people use higher proportions of polished rice.

In the evening, the soaked rice and black gram are ground separately with a mortar and pestle. For idli, the rice is coarsely ground and the black gram finely ground; for dosai, both are finely ground. Water, in a proportion twice the weight of rice and black gram, and 1 percent salt (weight to volume) are also added. The batter is incubated overnight. During this time the batter becomes acidic (about pH 4.5) and it is leavened by carbon dioxide produced by the principal fermenting microorganism *Leuconostoc mesenteroides*. *Streptococcus faecalis* is also present and contributes to the acid content.

The leavened, acidified batter is steamed in small cups to produce the idli cakes or fried like a pancake to yield dosai (Figure 2.1). Both are tasty, nutritious foods.

This process of producing wholesome protein-rich food can be adapted to other ingredients. Dehulled soybean can be substituted for black gram. Other starchy cereals could be substituted for the rice.

African Acidic Porridges

Naturally fermented acidic porridges are staple foods in many parts of Africa; for example, West African gari is prepared from an acid-fermented cassava porridge. During the fermentation, at an optimum temperature of 35°C, any cyanide-containing sugars present are hydrolyzed, removing the cyanide. The cassava becomes acidic, and the characteristic pleasing flavor of gari develops. The fermentation is usually complete in 3–4 days.

The principal microorganisms include *Corynebacterium manihot*, which hydrolyzes the starch, producing lactic and formic acids—a process evolving heat. As the product becomes more acidic (about pH 4.25), a yeast-like fungus *Geotrichum candidum* (also found in camembert cheese) develops, oxidizing the acid and producing the gari flavor.

Typically, the liquid is pressed from the ferment and the starchy residue is either used directly as fufu or dried in a basket over a fire with continual turning until it is converted to dry gelatinized granules, which can be stored for later consumption.

Related acidic porridges are made by fermentation of millet and maize (ogi), and mahewu (maize and wheat) and sorghum. The acidity protects the products from food-spoilage organisms, thus providing a wholesome food that keeps well in a relatively contaminated environment.

Limitations

The basic limitation to the introduction of these low-cost preservation technologies to new locations or countries is cultural preference and taste. This is particularly so in the case of fish, where, apart from the sauce and paste processes in Southeast Asia, most people will only eat "conventional" varieties. Fish, particularly in tropical conditions, spoils rapidly, and eating spoiled fish can have dangerous consequences. Canning and mechanical refrigeration are the only widely accepted methods of fish preservation added since ancient times, and both methods are too expensive for use by the poorest sections of developing countries.

Where the products of fermentation technologies are acceptable, however, there may be opportunities for increasing conservation of food resources through more widespread use of such methods, particularly with better quality control. There may be an important role here for the microbiologist in improving the techniques in ways appropriate to the local situation.

High salt content has excellent preservative action and is valuable as a condiment, but it restricts the amount of food that can be consumed. Lower salt contents, along with sufficient acid (pH 4.5), offer a satisfactory preservative action while permitting more consumption, thus contributing to better nutrition.

FIGURE 2.1 Steamed Indian idli cakes or same dough fried as a pancake (dosai). (Photograph courtesy of K. H. Steinkraus)

Generally, production of acidic cereal porridges like ogi and mahewu proceeds better at temperatures of about 50°C, which are favorable for rapid development of *Lactobacillus delbrueckii*. Fermentation time is shorter and other undesirable microorganisms have less chance of developing.

Research Needs

Research on the improvement and popularizing of fermented foods should be centered on:

- Studying acidic fermentations for their realiability in areas where they are not traditionally used; and
- Undertaking socioeconomic research to determine if consumers will accept the new products and, if not, how to encourage them to do so.

Improving Nutritional Value

Beers and wines make a valuable contribution to the proper nutrition of people subsisting on low incomes in the developing world. The fermentation processes involved raise the vitamin, protein, and, in some cases, the essential amino acid content of starchy substrates such as cassava, rice, maize, millet, sorghum, and other cereal grains.

Native maize or sorghum (kaffir) beers, indigenous rice wines, and alcoholic rice pastes bear little physical resemblance to Western beers and wines. The indigenous wines and beers are generally cloudy, opalescent, effervescent beverages, because of their content of microorganisms and substrate residues. All alcoholic beverages and foods provide a similar euphoria, depending on their alcoholic content, but Western beers and wines often provide the consumer with an excess of calories.

Most of the indigenous products provide essential nutrition to consumers in the form of vitamins, protein, amino acids, and calories. In addition to the food value in the basic ingredients, the microorganisms synthesize from these ingredients essential amino acids, protein, and vitamins that are consumed with the product. These organisms may also utilize a portion of the starch, reducing total solids and increasing the percentage of protein in the product, converting a low-protein food such as cassava to an acceptable staple in the diet.

A few of these fermentation processes will be described, and similar or related commodities can be produced wherever they might add valuable nutrition to the diet.

Two basic processes are used. The first involves germination (malting) of the grain, which produces enzymes (amylases) that transform a portion of the starch to sugars (glucose and maltose). The sugars are then fermented by yeasts such as *Saccharomyces cerevisiae* to ethyl alcohol. The yeast, however, also grows and synthesizes amino acids, proteins, and vitamins from the grain constituents. To retain all the nutrients in the beer or wine, it is essential that the products *not* be clarified or filtered.

The second process involves a starch-digesting mold (*Amylomyces rouxii*) and a yeast (*Endomycopsis burtonii*). The combination results in hydrolysis of the starch to sugars, which are then fermented to alcohol.

An example of the first process is kaffir (sorghum) beer. Kaffir beer is an alcoholic beverage with a pleasantly sour taste and the consistency of a thin gruel. It is the traditional beverage of the Bantu people of South Africa, and the alcohol content may vary from 1 to 8 percent. Kaffir beer is generally made from kaffircorn (*Sorghum caffrorum*), malt, and unmalted kaffircorn meal. Maize or millet (*Eleusine coracana*) may be substituted for part or all of the kaffircorn depending on the relative cost. Even cassava and plantains may be used, though with these the procedure would not be the same as with grain.

The kaffircorn grain is steeped for 6–36 hours. It is then drained and placed in layers and germinated by periodic moistening for 4–6 days. Germination continues until the bud is about 2.5 cm long and the material is then sun dried.

The essential steps are: mashing, souring, boiling, conversion, straining, and alcoholic fermentation.

Mashing is carried out in hot (50°C) water. Proportions of malted to unmalted grains vary, but 1 : 4 is satisfactory. Approximately 4 liters of water are added for about every 2 kg of grain. Souring begins immediately due to the presence of lactobacilli (*Lactobacillus delbrueckii),* at a temperature of 50°C.

Souring is complete in 6–15 hours. Water is added and the mixture is boiled. It is then cooled (to 40°-60°C) and more malt is added.

Conversion (starch hydrolysis) proceeds for 2 hours and then the mash is cooled (25°–30°C). Yeasts present in the malt are responsible for the natural fermentation, although *Saccharomyces cerevisiae* isolated from kaffir beer can be inoculated. Kaffir beer is ready for consumption in 4–8 hours, while it is still actively fermenting. The ethanol content is generally from 2 to 4 percent. The beer also contains from 0.3 to 0.6 percent lactic acid and from 4 to 10 percent solids. Production of acetic acid by *Acetobacter* species is the principal cause of spoilage.

Daily consumption of 3 liters of kaffir beer, made from approximately 0.5 kg of grain, is not unusual for a workingman. The improvement in the vitamin content of a diet that includes beer compared with a diet in which the kaffir corn is consumed directly is shown in Table 2.1.

TABLE 2.1 Comparison of Diet with and without Maize Beer

	Amount of Food Eaten (g)	
	Diet without Beer	Diet with Kaffir Beer
Food Item		
Maize, wholemeal	350	137.5
Maize, 60% extraction	350	137.5
Maize beer		5 pints (2840 ml)
Vegetables	130	130
Sweet potatoes	470	470
Kidney beans	30	30
Vitamin B_1	0.002	0.00195
Riboflavin	0.00113	0.00232
Nicotinic acid	0.0117	0.0203
Calories	3016	2979

Source: B. S. Platt. 1964. Biological ennoblement: improvement of the nutritive value of foods and dietary regimes by biological agencies. *Food Technology* (Chicago) 18:665.

The caloric content of the two diets is quite similar, only 37 calories being lost in the diet containing beer. The most notable improvement is the doubling of the riboflavin and the near doubling of nicotinic acid in the diet containing beer, because of synthesis of these vitamins during malting and fermentation. Pellagra, which is relatively common in people subsisting on maize, is never noted in those consuming usual amounts of kaffir beer.

An example of the second process is Indonesian tapé ketan, which is closely related to indigenous rice wine. It is a sweet-sour, alcoholic paste in which a starch-digesting mold (*Amylomyces rouxii*) and at least one yeast (*Endomycopsis burtonii*) hydrolyze steamed rice starch to maltose and glucose and then produce ethanol and organic acids, which provide an attractive flavor and aroma. Fermentation is complete in 2–3 days at 30°C. If yeasts of the genus *Hansenula* are present, the acids and ethanol are esterified, producing highly aromatic esters.

The inocula are obtainable in the markets of Indonesia as a product called ragi (in Thailand, luk-paeng). Ragi is a white dried-rice flour cake about 2.5 cm in diameter (Figure 2.2) containing a variety of molds and yeasts,

FIGURE 2.2 Indonesian ragi cakes used for inoculum for tapé ketan and tapé ketella. (Photograph courtesy of C. W. Hesseltine)

including those described above. Housewives prepare steam-soaked glutinous rice, inoculate it with the powdered ragi, place the inoculated rice into earthenware jugs with added water and allow the mixture to ferment for 3–5 days. The liquid portion is then consumed and additional water is added. Fermentation continues for 3–5 days, after which the liquid portion is again drunk. This is repeated until all the rice is fermented. Any residual dregs are sun dried and used as a type of ragi. Hence, there is no loss of nutrients.

Detailed studies have been made of the biochemical and nutritional changes that occur during tapé fermentations. Most of the starch is hydrolyzed to sugars, which, in turn, are fermented to ethanol and organic acids. Lysine, the main limiting amino acid in rice, is selectively synthesized by the microorganisms so that it increases by 15 percent. Thiamine, which is very low in polished rice (0.04 mg/100 g), is increased threefold (to 0.12 mg/100 g) by the action of the microorganisms. Up to 8 percent ethanol is produced; this serves as calories for consumers. Also, it probably contributes to destruction of disease-producing and food-spoiling organisms that might be present in the fermentation water.

Through the loss of total solids resulting from utilization of the starch, the protein content of tapé ketan is increased to as much as 16 percent (dry basis) compared with 7–8 percent in rice.

The tapé ketan process is a simple way of raising the protein quality in starch substrates and also of producing thiamine, which may be deficient in predominantly polished-rice diets.

Protein enhancement is all the more important in the case of tapé ketella, which is also produced in Indonesia. Tapé ketella is a sweet-sour alcoholic food made from cassava tubers. The tubers are peeled, steamed, and cut into pieces about 5 × 5 cm. They are then carefully inoculated on all surfaces with powdered ragi. A mold (*Amylomyces rouxii*) and yeasts of the genera *Endomycopsis* or *Hansenula*, along with related types, overgrow the cassava, utilizing a portion of the starch for energy. Cassava contains as little as 1 or 2 percent protein and by itself is clearly unable to contribute to proper human protein nutrition, even though it can provide sufficient calories. Consumption of a portion of the cassava as tapé ketella, which may contain 8 percent protein, can have a beneficial effect on nutrition.

Limitations

In establishing these fermentations in areas of the world where they are unknown, the proper cultures should be obtained from either culture collections or scientists who have done research on the products. Acceptance of new foods will be more difficult. Technical studies must be accompanied by socioeconomic studies of the potential role of these products in the particular society.

Research Needs

Small-scale laboratory studies are needed in applying these processes in a new environment.

Production of Meat-Like Flavors

Shoyu (soy sauce) and miso (soybean paste) are made by similar processes from soybeans by using koji prepared with the molds *Aspergillus soyae* and *A. oryzae.* The koji process for culturing microorganisms for various purposes is described below. Both products are salty and are used to add flavor to vegetables, fish, and meat. Miso exists in many colors and flavors and is a paste, whereas shoyu is a liquid from which the insoluble solids have been removed.

Koji is generally made from rice, although some forms involve the use of barley or soybeans. The finished koji is added to soaked, pressure-cooked, whole soybeans. When these products are mashed together, a salt-tolerant yeast, *Saccharomyces rouxii*, and considerable amounts of salt (4–13 percent weight to weight) are added. The salt is added for flavor and to retard growth of toxin-forming bacteria. The mash is then placed in tanks made of concrete or wood and containing several tons of substrate. The mash in the tanks is allowed to ferment from a few days to a number of months, depending on the type of miso desired. On completion of the fermentation, miso is either ground into a uniform paste with the consistency of peanut butter or packaged directly.

Traditionally, miso is used as a flavoring base for soup eaten at breakfast. To this base, vegetables and seafood are added. Miso imparts a meat-like flavor to the soup and is added to fish and meats before baking or broiling. Currently, it is also being incorporated into sauces for pizza and spaghetti and is used as an ingredient in some commercially prepared salad dressings.

Shoyu manufacture differs from miso in that wheat that has been cleaned, roasted, and crushed is used in place of rice to make the koji. Whereas miso requires the use of whole soybeans, modern shoyu manufacture utilizes defatted soybean flakes, which are moistened and blended in a ratio of 55 percent soybean flakes to 45 percent crushed wheat. This mixture of wheat and soybeans is inoculated with selected strains of a mold, *Aspergillus oryzae* or *A. soyae*, and transformed to koji as it becomes overgrown with mold (Figure 2.3). During molding the temperature is held below 40°C, and about 3–4 days are required for completion of the process. At this time, about an equal amount of brine is added to the koji and the mixture is placed in large tanks and inoculated with a yeast, *Saccharomyces rouxii*, and a bacterium of the *Lactobacillus* species. Depending on the temperature, the mash is allowed

FIGURE 2.3 First stage in traditional soybean fermentation. (Photograph courtesy of K. H. Steinkraus)

to ferment without aeration for 6–9 months. When this fermentation is completed, the material is pumped to presses where the dark-brown liquid is pressed out, pasteurized, and bottled. The press cake is used for cattle feed.

Shoyu contains a large amount of glutamic acid, salt at the level of about 18 g per 100 ml of liquid, reducing sugars, and alcohol. The nitrogen compounds consist of about 40–50 percent amino acids, 40–50 percent peptides, and less than 1 percent protein. In the Chinese soy sauce process more soybeans are added and less wheat. A chemical process is also used in which the soybeans are hydrolyzed with acid. This is an inferior process in that the product is harsh to the taste and is generally blended with fermented shoyu, or the flavor is modified by the addition of various flavoring agents.

Because of the high salt content, pasteurization, and the addition of preservatives, soy sauce can be kept for months without refrigeration. Similar products can be made using barley, coconut, and even hydrolyzed yeast cells. Shoyu is widely used both in the Orient and in Western countries. A modern shoyu fermentation plant has been operating in the United States for several years.

Limitations

The basic limitation of shoyu and miso-type products is shared with other food products—cultural resistance to unconventional foods. In the case of shoyu and miso, this is compounded by the difficulty of growing soybeans in tropical countries and by their shortage or high price. The possibility of using other, more readily available, legumes* should be investigated.

The process of making miso and soy sauce generally takes considerable time. The technology is complicated and requires considerable experience and training before an acceptable product can be produced. Although it can be produced at the village level, manufacture on a large scale is more efficient and usually yields a better and more uniform product.

Extreme care must be taken in selecting industrial strains of molds to ensure that they are *Aspergillus oryzae* and not a closely related species, *A. flavus,* which produces the highly toxic aflatoxin.

Research Needs

Research should be concentrated on:

- Evaluating the potential of miso as a meat-flavoring agent in the local preparation of foods;
- Shortening fermentation time to reduce salt content to a minimum level;
- Evaluating the potential of other legumes and cereals such as sorghum, millet, and corn to replace soybeans, rice, and wheat (research indicates that corn might be used to make miso koji); and
- Replacing the salt with another bacteriostat to make the products palatable for children, or using dehydration, boiling, and canning.

Koji Method of Producing Enzymes

Koji is the Japanese name for the solid-state culture of microorganisms on rice, barley, wheat, soybeans, and other cereals. The microorganisms used are the typical molds (*Aspergillus oryzae*, *A. soyae*, and species of *Rhizopus* and *Monascus*).

The substrate is soaked in water, drained, heat-sterilized, cooled, and then inoculated with spores of the appropriate koji mold. The inoculated substrate is then placed in trays or in large shallow tanks in an incubator room. An incubator is not always required. The mold is allowed to grow for 2 or 3 days, with occasional turning of the material either mechanically or by hand. At the end of the fermentation, the moist molded material is an excellent source of enzymes.

Koji is used as a source of enzymes in the manufacture of shoyu, miso, and sake (rice wine). For each of the above foods, special mold strains are used,

*See also National Academy of Sciences, *The Winged Bean: A High-Protein Crop for the Tropics*, 1975, and *Tropical Legumes: Resources for the Future*, 1979, Washington, D.C.

and often two or three strains are combined. For example, one mold will produce the desired enzyme for breaking down starch to sugar and another will produce enzymes that hydrolyze proteins. Thus, the koji inoculum for making sake will not necessarily be the same as that used for making soy sauce.

The koji process can be modified by using different substrates and various molds to produce enzyme preparations used in different industrial processes. For example, a koji process is used to manufacture microbial rennet to make curd in the cheese industry. It has also been used successfully to produce coloring agents and in the continuous production of feed made from liquid animal waste mixed with maize. A process has been reported that uses the soybean residue from soybean milk manufacture to make a tempeh-like food product (to be discussed in the next section).

The koji process has the following advantages:

- The fermentation equipment can be as simple as, for example, wooden trays.
- Fermentation substrates can be low-cost. For instance, only broken and damaged rice is used to make koji for the sake fermentation.
- Energy requirements are low, since forced aeration is not required and once the mold starts to grow it usually produces sufficient heat to warm the air in the fermentation room to the proper temperature.
- After inoculation, a pure culture fermentation is unnecessary.
- Extraction of a product is simple, since it is not necessary to extract a large volume of liquid, which could create a pollution problem after product recovery.
- Yields may be much greater with solid substrates than in liquid media.
- Presumably, any product produced by fungi can be made by this process after a suitable mold culture is found.

Limitations

The inoculum must be produced in large quantities and be as free of contamination as possible.

The koji molds often generate excess heat, and cooling with fans is necessary for maximum product formation.

The moisture and temperature of the fermenting substrate must be controlled at the optimum levels required by the mold to produce the desired product efficiently. Since molds require air for growth, the molding material must be turned regularly and should not be too deep or dense. Thus, the process requires trained personnel to ensure quality control in a nontraditional production process.

Research Needs

Research should be concentrated on:

- Improving molds for making koji by selection and mutation to produce more of the product in a shorter period and to reduce the requirement for aeration. Any enzyme used in the food industry can be manufactured by this process if a suitable mold capable of producing the enzyme is available.
- Testing in animals any fungus selected for food production to be sure it does not form toxin.
- Converting batch-type fermentation to continuous fermentation.

Indonesian Tempeh

Tempeh, a vegetarian meat analogue and source of vitamin B-12 (generally lacking in vegetarian diets), is a product made in the East Indies by fermenting soaked, partially cooked, dehulled soybeans (Figure 2.4).

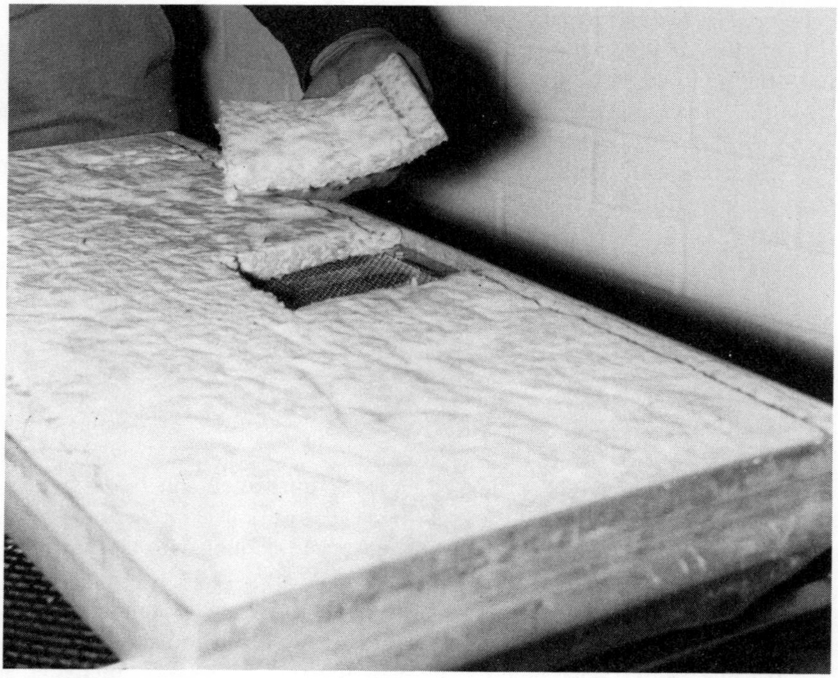

FIGURE 2.4 Tempeh soybean cake: the soybeans are covered and knitted into a cake by the mold mycelium. (Photograph courtesy of K. H. Steinkraus)

In the tropics, the fungal fermentation with *Rhizopus* spp. is preceded by a bacterial acid fermentation during soaking of the soybeans. This increases the acidity of the beans (pH 5.0), which is favorable to subsequent growth of the mold but inhibits many bacteria that could cause spoilage of the tempeh. In temperate climates, the bacterial fermentation does not readily occur and some researchers advise acidification of the beans during cooking by the addition of 1 percent lactic acid or 0.5 percent acetic acid.

Optimum temperature for the fermentation is between 30°C and 37°C. At such temperatures, fungal growth is completed in 24 hours or less. At temperatures below 30°C, the fermentation may require 2 or 3 days. Fermentation is complete when the beans are knitted into a compact cake by the mold mycelium. The cake can then be sliced thin and fried in deep fat, or cut into chunks and used in soups as a meat replacement.

The fresh tempeh has a pleasant, dough-like aroma. Tempeh is frequently dipped in shrimp, fish, or soy sauces prior to or after deep-fat frying. The taste of fried tempeh is bland, with a slightly nutty flavor acceptable to nearly everyone.

Interestingly, it has been found that commercial tempeh shows vitamin B-12 activity, produced by an unidentified bacterium that grows on the soybeans simultaneously with the mold. If the tempeh is made by inoculating with pure mold, no vitamin B-12 is synthesized. Thus, tempeh prepared with the mold and the bacterium can provide not only protein but also vitamin B-12. This is extremely important for vegetarians, whose diet might otherwise be deficient in this important vitamin.

In Indonesia soybeans are fermented in packets made from wilted banana leaves, with the mold inoculum coming from a previous batch of uncooked tempeh (Figures 2.5 and 2.6). A similar product, ontjom (oncom), is made in the same region as tempeh, but with peanut press cake as the substrate and *Neurospora sitophila* as the fermenting fungus. Peanut press cake is the residue left after peanut oil has been removed, and it is high in protein. The fermentation technology is similar to that for tempeh and the end product is a pinkish-textured meat substitute with an almond or mincemeat flavor. It is prepared in small factories (cottage industries) and eaten after deep-fat frying or in soups.

Tempeh has been made using various cereals such as wheat, sometimes in combination with soybeans, to make a product that tastes like bread or popcorn. Tempeh, as well as some other fermented foods, has been shown to contain antibiotic substances active against certain types of bacteria.

Research conducted on the nutritional value of tempeh shows that it is a wholesome, nutritious food. It contains 42 percent protein, derived from the soybeans and from the protein synthesized by the mold. Neither the tempeh nor the ontjom fungus is known to produce any toxins, and they have been

FIGURE 2.5 Small packets of tempeh as sold in Indonesian markets. Wilted banana leaves are used to cover the dehulled soybeans during fermentation. (Photograph courtesy of K. H. Steinkraus)

consumed in the East Indies for hundreds of years. Some of the B-vitamins such as riboflavin, B-12, and niacin are increased during the fermentation. Feeding studies using rats indicate that the protein efficiency ratio (PER) values are similar to those of soybeans.

Potentially, tempeh can be produced in many parts of the world without elaborate equipment or extensive training. Since the food is bland, it can be modified to suit local tastes by adding appropriate sauces and spices.

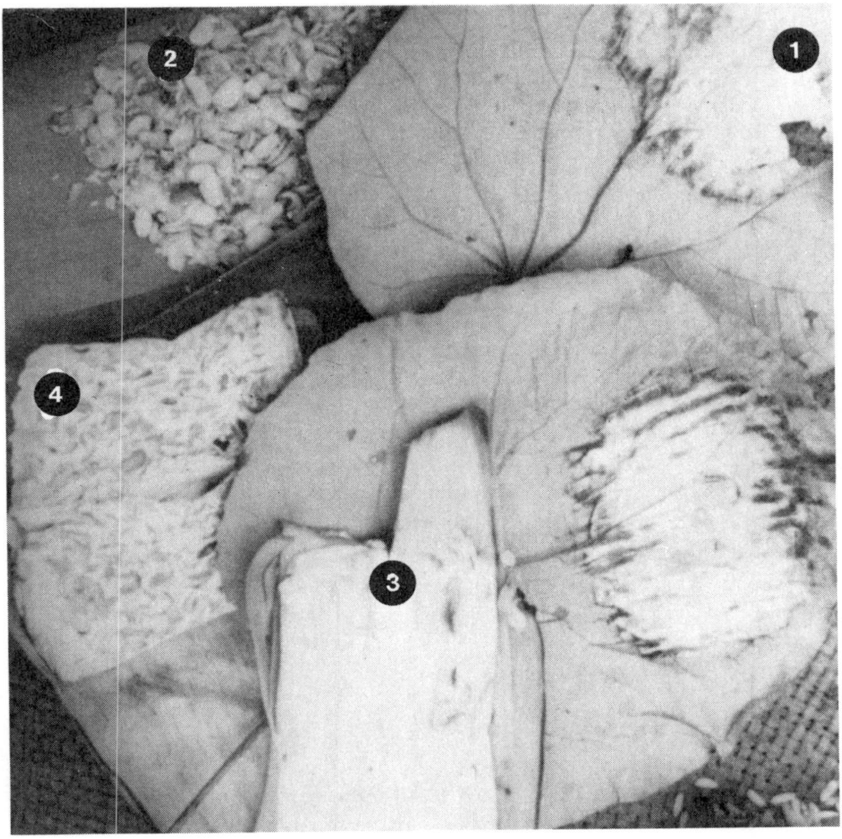

FIGURE 2.6 1. Tempeh mold inoculum grown on leaf. 2. Dehulled, partially cooked soybeans. 3. Tempeh cake. 4. Sliced tempeh cake. (Photograph courtesy of K. H. Steinkraus)

Limitations

Tempeh must be eaten within a day or two after fermentation unless it is dried or steamed and refrigerated. When dehydrated, it will keep for months.

The technology is based on village-level processes, and large-scale equipment for its production has not yet been developed.

It is important that only known and proven cultures of the mold be used. Wild strains of the mold should not be used because they may contain a toxin or may not produce an acceptable flavor.

Bacterial contamination and spoilage during fermentation can be a problem if the cooked beans are too moist or if they have not been acidified.

In general, tempeh is subject to the same limitations as miso and shoyu.

FLOW SHEET: Indonesian Tempeh Fermentation

Whole, clean soybeans

↓

Soak overnight (or 24 hours) to hydrate soybeans and allow bacterial fermentation and acidification*

↓

Dehull by hand or by passing through machine to loosen hulls

↓

Remove hulls by flotation on water

↓

Boil cotyledons for 60 minutes

↓

Drain and cool and allow surface moisture to evaporate

↓

Inoculate with tempeh mold

↓

Ferment small packets of inoculated soybeans wrapped in banana leaves or in clean shallow covered pans

↓

Incubate at a temperature of 30° - 35°C until soybeans are completely covered with mold mycelium (generally 24 to 36 hours)

↓

Tempeh cakes can be sold on the market or used in home by slicing thin strips and deep frying or cutting into chunks and cooking in soups

Research Needs

Research requirements include:

- Establishing a small company or laboratory with microbiological know-how to produce a dry, pure tempeh culture inoculum in small packages for distribution at low cost. This is important for the production of tempeh in small cottage industries that do not have technically trained people.
- Standardizing the inoculum, as has been done with baker's yeast, to yield a product that will produce a uniform tempeh under standard conditions of time and temperature. The inoculum should have good keeping properties and should not require refrigeration.
- Conducting research on the control of mold contamination in rural environments.
- Adapting simple fermentation equipment for tempeh production at the local,level in countries unfamiliar with its production.
- Studying sociocultural aspects of introducing tempeh and tempeh-like products to people unfamiliar with the product or with soybeans.

*In a temperate climate it may be necessary to add 0.5 percent vinegar during the cooking to increase the acidity.

Single-Cell Protein Production

Single-cell protein (SCP) refers to the cells of yeasts, bacteria, fungi, and algae grown for their protein content. Cells of these microorganisms also contain carbohydrates, fats, vitamins, and minerals. SCP products are used either for animal feed or human food. They are potentially very important sources of amino acids, proteins, vitamins, and minerals that can be prepared from otherwise inedible or low-quality waste material.

Yeasts

Yeasts in baked and fermented food products have a long history of human consumption. Dried brewer's yeast, a by-product of the brewing industry, has an established use in animal feed formulations. It is also used in human nutrition as a "health food" dietary supplement.

In recent years SCP processes have been practiced on a commercial scale, based on the growth of yeasts in deep-tank, agitated, and aerated cultures. Examples of raw materials used in these processes and yeasts that utilize them are molasses, *Saccharomyces cerevisiae*; n-paraffin hydrocarbons, *Saccharomycopsis lipolytica;* and cheese whey, *Kluyveromyces fragilis.* The Symba process, developed in Sweden, utilizes starchy wastes by combining two yeasts, *Saccharomycopsis fibuligera* and *Candida utilis.*

Bacteria

Large-scale propagation of bacteria as a source of animal feed protein has been considered only in the last decade. A large-scale (75,000 t) facility for producing the methanol-utilizing bacterium *Methylophilus methylotrophus* is being constructed in the United Kingdom, and a large pilot plant for growing a stable mixed culture of methane-utilizing bacteria is being operated in The Netherlands.

The conversion of cellulosic materials such as bagasse from sugar cane processing to SCP by bacteria of the genera *Cellulomonas* and *Alcaligenes* has been investigated on a laboratory and small pilot-plant scale at Louisiana State University. Plans call for commercial-scale production. (The microbiological utilization of cellulose is discussed in Chapter 8 of this report.)

Advantages claimed for bacteria over yeasts for production of SCP include more rapid generation and a higher content of crude protein and certain essential amino acids, particularly methionine. However, bacterial cells are smaller than yeasts and may be more costly to harvest unless the cells can be flocculated to give a higher solids slurry prior to centrifugation. Further (apart from fermented milks and cheeses, which often contain as many bacteria as 5×10^9 per g of foodstuff), bacteria as such have had only a brief history of use as either animal feed or human food. Food and drug regulatory agencies in most countries will have to be convinced of the safety of bacterial products before permitting their use.

Fungi

People have eaten higher fungi, particularly mushrooms, since ancient times. Recently, a different method of growing fungal mycelium, including mushroom mycelium, has been developed on a large pilot-plant or commercial scale in deep-tank, aerated, and agitated cultures. Typical raw materials and organisms are shown in Table 2.2.

Problems encountered with production of fungal mycelium as a source of SCP include slow growth rates and the consequent need to maintain sterile conditions over an extended time to prevent overgrowth by bacterial and wild yeast contaminants. This increases costs for fungal mycelium production. In recent laboratory-scale studies, *Chaetomium cellulolyticum*, a thermo-tolerant cellulolytic fungus, has shown promise in the conversion of cellulose to SCP.

Care must be taken to use strains of fungi that do not produce mycotoxins that affect domestic livestock or human beings.

Algae

Algae are of interest as a source of SCP because they grow well in open ponds and utilize carbon dioxide as a carbon source and sunlight as an energy source for photosynthesis. Algae of the genera *Chlorella* and *Scenedesmus* have been grown for food use in Japan. *Spirulina* species have been eaten for many years by inhabitants of the northern shores of Lake Chad in Africa and by the Aztec Indians in Mexico, where they are now being grown on a pilot-plant scale in the alkaline waters of Lake Texcoco. *Spirulina* is a particularly

TABLE 2.2 Raw Materials Used in Growing Fungi Commercially

Raw Materials	Fungal Species
Cane and beet molasses	*Agaricus campestris*
Maize syrups, dexture (D-Glucose), cheese whey, and canning wastes	*Morchella esculenta, M. hortensis, M. crassipes* (Morel mushrooms)
Coffee-processing wastes	*Trichoderma* spp.
Maize wet-milling waste	*Gliocladium* spp.
Maize and pea-canning wastes	*Trichoderma reesei*

attractive algal source of SCP because of its high nutritive value and because the large cell filaments make it relatively easy to harvest by fine mesh nets or filtration. Further discussions of algae are in the section on wastes in Chapter 7 and in the NAS publication *Underexploited Tropical Plants with Promising Economic Value.*

Limitations

SCP production is capital-intensive and, with the exception of algal production by photosynthesis, energy-intensive. Processes that must be conducted under sterile conditions require stainless-steel equipment that can be cleaned and sterilized. They also require provisions for sterilizing the growth medium and recovering the SCP product without introducing other microbial contaminants, particularly human pathogens. Trained personnel are needed to supervise and maintain quality control of production.

At present, SCP processes for the production of animal feed are the most attractive, since conventional animal feedstuffs such as soybean meal and fish meal must be imported to many tropical and subtropical countries at prevailing international prices. In using SCP for animal feed, however, there is a large loss of conversion efficiency as opposed to direct human use. For human food applications, the use of microorganisms is limited to those such as *Saccharomyces cerevisiae* and *Candida utilis* that are accepted by regulatory and public health authorities as safe for human food use. Even those organisms, if they are to be consumed as a significant portion of the protein in the diet, must be processed further to reduce nucleic acid contents to below the levels that could lead to kidney stone formation or gout.

To achieve economies of scale, an SCP production facility should have a capacity of at least 50,000 t per year unless operated as a waste-treatment facility in a food-processing plant. This implies that sufficient raw materials will be available in close proximity to the SCP plant to meet these production requirements.

For raw materials, carbohydrates such as sugars or sugar-containing by-products, wastes, and starches are likely to be available in semitropical and tropical countries in the quantities required for an economically viable scale of production. In general, concentrations of utilizable carbohydrates in wastes should be sufficiently high that handling of large volumes of dilute materials is avoided.

A portion of the crude molasses produced from sugar cane operations could be diverted to SCP production (yeasts), if a source of nitrogen were added to provide a source of protein and vitamins for animal feeds, particularly poultry rations.

Many cities in tropical and semitropical countries have breweries that produce residues of reasonably uniform composition throughout the year. In addition to recovery of brewer's yeast, already widely used as animal feed, the remaining liquid waste—after hydrolysis to simple sugars—can be used as a potential carbon and energy source for SCP production. However, carbon-to-nitrogen ratios of the material may have to be adjusted to favorable ranges for yeast growth. In the case of starchy crops such as cassava, large quantities must be available at one site to provide a sufficient source of raw material for economic SCP production.

Coffee-processing wastes contain soluble carbohydrates and have a high chemical oxygen demand (COD) and soluble solids content. Pilot-scale operations in Guatemala have shown that growing *Trichoderma* species in these wastes reduces the COD considerably and yields an SCP product of interest for use in animal feed.

Microorganisms require sources of nitrogen, phosphorus, and mineral salts for growth, in addition to a carbon and energy source. The availability of ammonium salts such as ammonium sulfate or diammonium phosphate may be a problem in some countries. The same can be said for sources of phosphorus. A feed-grade source of phosphoric acid or soluble phosphates should be used because of the presence of arsenic and fluoride in crude phosphates. Other minerals are usually present in the water supply.

To reduce contamination to a low level, the microorganism used should multiply (grow) rapidly at an acid level of pH 4.5 or below. Operations under these conditions will allow the use of clean, aseptic conditions without the need for sterile facilities. SCP production (except from algae) requires aeration to achieve suitable yields. Air should be filtered to remove contaminants. Power costs for aeration, fluid handling, and steam for cleaning, recovery, and drying the product can be significant factors in the total energy costs.

Water requirements for SCP production are considerable for both processing and cooling. The growth of microorganisms produces heat, which must be removed to maintain the growth temperature within the preferred range of 30°-35°C. Cooling water temperatures should be at least 10°C lower than the growth temperature. Location of an SCP plant near the ocean would permit the use of seawater for cooling. Also, wastewater is produced from SCP operations and must either be disposed of or, preferably, treated and recycled in large operations.

Three SCP processes appear to have potential for further development. These processes, although known, are not used in tropical and semitropical regions to the extent that they might be. They are given in Table 2.3.

All of these processes can be carried out in either a batch or continuous mode of operation. They should be operated under clean, aseptic conditions, but they do not require tight control over sterility throughout the process.

TABLE 2.3 Production of SCP from Various Substrates

Substrate	Organisms	Conditions
Cane molasses	*Candida utilis* *C. tropicalis* *Rhodotorula gracilis* *R. pilimanae* *R. rubra*	Temperature: 30°–34° C pH 4.0–4.5
Coffee wastes	*Candida* spp. *Trichoderma* spp.	Temperature: 30°–35°C pH 2.0–4.0
Starchy materials, especially cassava	Symba processes: mixed culture of *Saccharomycopsis fibuligera* and *Candida utilis*	Temperature 30°–34°C pH 4.0–5.0

Research Needs

In many countries there is need to:

- Conduct feasibility studies to determine where there is an appropriate mix of underutilized residue, technical competence, and need or economic opportunity to use the SCP produced;
- Conduct studies on the use of yeasts, e.g., genus *Rhodotorula*, which have relatively rapid growth rates (2- to 2.5-hour generation times), temperature over the range 28° to 34°C and pH tolerance between 3.5 and 5.0;
- Institute animal feeding studies using dried SCP as a component of poultry and swine rations;
- Investigate a wider range of thermotolerant organisms, particularly yeasts, for their utility in producing SCP from coffee-processing wastes;
- Evaluate nutritional and safety factors for animal feed applications of certain thermotolerant fungi including *Sporotrichum thermophile* and *Paecilomyces* species that grow on cassava; and
- Develop thermotolerant strains of microorganisms to reduce the requirement for cooling water in semitropical and tropical regions.

References and Suggested Reading

Food Preservation

Akinrele, I. A. 1963. *Further studies on the fermentation of cassava.* Research Report No. 20. Lagos, Nigeria: Federal Institute of Industrial Research.

Morcos, S. R.; Hegasi, S. M.; and el-Damhougy, S. T. 1973. Fermented foods in common use in Egypt. I. The nutritive value of kishk. *Journal of the Science of Food and Agriculture* 24:1153-1156.

Mukerjee, S. K.; Albury, M. N.; Pederson, C. S.; van Veen, A. G.; and Steinkraus, K. H. 1965. Role of *Leuconostoc mesenteroides* in leavening the batter of idli, a fermented food of India. *Applied Microbiology* 13:227-231.

Pederson, C. S., and Albury, M. N. 1969. *The Sauerkraut fermentation.* New York Agricultural Experiment Station Technical Bulletin 824. Geneva, New York: New York State Agricultural Experiment Station.

Schweigert, F.; Van Berge, W. E. L.; Wiechers, S. G.; and de Wit, J. P. 1960. *The production of mahewu.* Report No. 167. Pretoria, South Africa: Council for Science and Industrial Research.

Stamer, J. R. 1975. Recent developments in the fermentation of sauerkraut. In *Lactic acid bacteria in beverages and food*, J. G. Carr; C. V. Cutting; and C. S. Whiting, eds., pp. 267-280. New York: Academic Press.

Steinkraus, K. H.; van Veen, A. G.; and Thiebeau, D. P. 1967. Studies on idli–an Indian fermented black gram-rice food. *Food Technology* (Chicago) 21:916-919.

van Veen, A. G. 1965. Fermented and dried seafood products in Southeast Asia. In *Fish as Food*, G. Borgstrom, ed., Volume 3, pp. 227-250. New York: Academic Press.

_______ ; Hackler, L. R.; Steinkraus, K. H.; and Mukerjee, S. K. 1967. Nutritive value of idli, a fermented food of India. *Journal of Food Science* 32:339-341.

Improving Nutritional Value

Cronk, T. C.; Steinkraus, K. H.; Hackler, L. R.; and Mattick, L. R. 1977. Indonesian tapé ketan fermentation. *Applied Environmental Microbiology* 33:1067-1073.

Ellis, J. J.; Rhodes, L. J.; and Hesseltine, C. W. 1976. The genus *Amylomyces. Mycologia* 68:131-143.

Ko, S. D. 1972. Tapé fermentation. *Journal of Applied Microbiology* 23:976-978.

Novellie, L. 1968. Kaffir beer brewing: ancient art and modern industry. *Wallerstein Laboratories Communications* 31:17-32.

Platt, B. S. 1946. Fermentation and human nutrition. *Proceedings of the Nutrition Society* 4:132-140.

____. 1955. Some traditional alcoholic beverages and their importance in indigenous African communities. *Proceedings of the Nutrition Society* 14:115-124.

____. 1964. Biological ennoblement: improvement of the nutritive value of foods and dietary regimes by biological agencies. *Food Technology* (Chicago) 18:662-670.

____; and Webb, R. A. February 1948. Microbiological protein and human nutrition. *Chemistry and Industry* 7:88-90.

Schwartz, H. M. 1956. Kaffircorn malting and brewing studies. I. The kaffir beer brewing industry in South America. *Journal of the Science of Food and Agriculture* 7:101-105.

Production of Meat-Like Flavors

Ebine, H. 1972. Miso. In *Proceedings of the International Symposium on Conversion and Manufacture of Foodstuffs by Microorganisms*, pp. 127-139. Tokyo: Saikon Publishing Company.

Hesseltine, C. W., and Shibasaki, K. 1961. Miso III. Pure culture fermentation with *Saccharomyces rouxii. Applied Microbiology* 9:515-518.

____, and Wang, H. L. 1967. Traditional fermented foods. *Biotechnology and Bioengineering* 9:275-288.

National Academy of Sciences. 1975. *The winged bean: a high-protein crop for the tropics.* Report of an *Ad Hoc* Panel of the Advisory Committee on Technology Innovation, of the Board on Science and Technology for International Development. Washington, D.C.: National Academy of Sciences.

____. 1979. *Tropical legumes: resources for the future.* Report of an *Ad Hoc* Panel of the Advisory Committee on Technology Innovation, of the Board on Science and Technology for International Development. Washington, D.C.: National Academy of Sciences.

Shibasaki, K., and Hesseltine, C. W. 1962. Miso-fermentation. *Economic Botany* 16:180-195.
Shurtleff, W., and Aoyagi, A. 1977. *The book of miso.* Brookline, Massachusetts: Autumn Press.
Yokotsuka, T. 1960. Aroma and flavor of Japanese soy sauce. *Advances in Food Research* 10:75-134.
_____. 1972. Shoyu. In *Proceedings of the International Symposium on Conversion and Manufacture of Foodstuffs by Microorganisms*, pp. 117-125. Tokyo: Saikon Publishing Company.

Koji Method of Producing Enzymes

Hesseltime, C. W. 1972. Solid-state fermentations. *Biotechnology and Bioengineering* 14:517-532.
_____; Swain, E. W.; and Wang, H. L. 1976. Production of fungal spores as inoculum for oriental fermented foods. *Developments in Industrial Microbiology* 17:101-115.
Nakano, M. 1972. Synopsis on the Japanese traditional fermented foodstuffs. In *Waste recovery by microorganisms*, pp. 27-28. Kuala Lumpur: United Nations Educational, Scientific, and Cultural Organization, distributed in the United States by UNIPUB, New York.
Sakaguchi, K. 1972. Development of industrial microbiology in Japan. In *Proceedings of the International Symposium on Conversion and Manufacture of Foodstuffs by Microorganisms*, pp. 7-10. Tokyo: Saikon Publishing Company.

Indonesian Tempeh

Hesseltine, C. W. 1965. A millennium of fungi, food and fermentation. *Mycologia* 57:149-197.
Liem, I. T. H.; Steinkraus, K. H.; and Cronk, T. C. 1978. Production of vitamin B-12 in tempeh—a fermented soybean food. *Applied and Environmental Microbiology* 34:773-776.
Roelofsen, P. A., and Talens, A. 1964. Changes in some B vitamins during molding of soybeans by *Rhizopus oryzae* in the production of tempeh kedelee. *Journal of Food Science* 29:224-226.
Steinkraus, K. H.; Bwee Hwa, Y.; Van Buren, J. P.; Provvidenti, M. I.; and Hand, D. B. 1960. Studies on tempeh—an Indonesian fermented soybean food. *Food Research* 26:777-788.
_____; Van Buren, J. P.; Hackler, L. R.; and Hand, D. B. 1965. A pilot-plant process for the production of dehydrated tempeh. *Food Technology* (Chicago) 19:63-68.
van Veen, A. G.; Graham, D. C. W.; and Steinkraus, K. H. 1968. Fermented peanut press cake. *Cereal Science Today* 13:96-99.
Wang, H. L.; Swain, E. W.; and Hesseltine, C. W. 1975. Mass production of *Rhizopus oligosporus* spores and their application in tempeh fermentation. *Journal of Food Science* 40:168-170.

Single-Cell Protein Production

Aguirre, F.; Moldonado, O.; Rolz, C.; Menche, J. F.; Espinosa, R.; and Cabrera, S. 1976. Protein from waste-growing fungi on coffee waste. *Chemical Technology* 6:636-642.
Baens-Arcega, L. 1969. Philippine contribution to the utilization of microorganisms for the production of foods. In *Biotechnology and Engineering Symposium*, Second International Conference on Global Impacts of Applied Microbiology, Addis Ababa, Ethiopia, Elmer L. Gaden, Jr., ed., pp. 53-62. New York: John Wiley and Sons.
Brook, E. J.; Stanton, W. K.; and Wallbridge, A. 1969. Fermentation methods for protein enrichment of cassava. *Biotechnology and Bioengineering* 11:1271-1284.
Davis, P., ed. 1974. *Single-cell protein.* New York: Academic Press.
Khor, G. L.; Alexander, J. C.; Santos-Nunez, J.; Reade, A. E.; and Gregory, K. F. 1976. Nutritive value of thermotolerant fungi grown on cassava. *Canadian Institute of Food Science and Technology Journal* 9:139-216.

Lewis, C. W. 1976. Energy requirements for single-cell protein production. *Journal of Applied Chemistry and Biotechnology* 26:568-576.

Litchfield, J. H. 1974. The facts about food from unconventional sources. *Chemical Processing* (London) 20:11-18.

_____. 1977. Single-cell proteins. *Food Technology* (Chicago) 31:175-179.

Mateles, R. I., and Tannenbaum, S. R., eds. 1968. *Single-cell protein.* Cambridge, Massachusetts: Massachusetts Institute of Technology Press.

Moo-Young, M. 1977. Economics of SCP production. *Process Biochemistry* (London) 12:6-10.

_____; Chahal, D. S.; Swan, J. E.; and Robinson, C. W. 1977. SCP production by *Chaetomium cellulolyticum*, a new thermotolerant cellulolytic fungus. *Biotechnology and Bioengineering* 19:527-538.

National Academy of Sciences. 1975. *Underexploited tropical plants with promising economic value*. Report of an *Ad Hoc* Panel of the Advisory Committee on Technology Innovation, Board on Science and Technology for International Development, Commission on International Relations. Washington, D.C.

Peppler, H. J., ed. 1978. *Microbial technology.* New York: Krieger Publishing Company.

Ratledge, C. 1975. The economics of single-cell protein production. *Chemistry and Industry* (London) 21:918-920.

Richmond, A., and Vonshak, A. 1978. Algae—an alternative source of protein and biomass for arid zones. *Arid Lands Newsletter* 9:1-7.

Storasser, J.; Abbott, J. A.; and Battey R. F. 1970. Process enriches cassava with protein. *Food Engineering*, May: 112-116.

Tannenbaum, S. R., and Wang, D. I. C., eds. 1975. *Single-cell protein II.* Cambridge, Massachusetts: Massachusetts Institute of Technology Press.

Waslien, C. I. 1975. Unusual sources of protein for man. *CRC Critical Reviews in Food Science and Nutrition* 5:77-151.

Wiken, T. O. 1972. Utilization of agricultural and industrial wastes by utilization of yeasts. In *Proceedings of the Fourth International Fermentation Symposium, Kyoto, Japan*, pp. 569-576. Osaka: Society of Fermentation Technology.

Sources of Cultures

Food Preservation

American Type Culture Collection, 12301 Parklawn Drive, Rockville, Maryland 20852, U.S.A.

Improving Nutritional Value

American Type Culture Collection, 12301 Parklawn Drive, Rockville, Maryland 20852, U.S.A.

Centraalbureau voor Schimmelcultures, P.O. Box 273, 3740 AG, Baarn, The Netherlands.

National Collection of Yeast Cultures, Lyttell Hall, Nutfield, Redhill, Surrey RH1 4HY, England.

Production of Meat-Like Flavors

American Type Culture Collection, 12301 Parklawn Drive, Rockville, Maryland 20852, U.S.A.

Centraalbureau voor Schimmelcultures, P.O. Box 273, 3740 AG, Baarn, The Netherlands.

Koji Method of Producing Enzymes

American Type Culture Collection, 12301 Parklawn Drive, Rockville, Maryland 20852, U.S.A.

Centraalbureau voor Schimmelcultures, P.O. Box 273, 3740 AG, Baarn, The Netherlands.

Indonesian Tempeh

Centraalbureau voor Schimmelcultures, P.O. Box 273, 3740 AG, Baarn, The Netherlands.

American Type Culture Collection, 12301 Parklawn Drive, Rockville, Maryland 20852, U.S.A.

Single-cell Protein Production

American Type Culture Collection, 12301 Parklawn Drive, Rockville, Maryland 20852, U.S.A.

Research Contacts

Food Preservation

O. Kandler, Botanisches Institut, Der Universität München, 800 München 19, Menzinger Strasse 67, Federal Republic of Germany.

Carl S. Pederson, Professor of Microbiology, *Emeritus.* Cornell University, Geneva, New York 14456, U.S.A.

Keith H. Steinkraus, Department of Food Science and Technology, New York State Agricultural Experiment Station, Geneva, New York 14456, U.S.A.

Reese Vaughn, Professor Emeritus, Department of Food Science and Technology, University of California, Davis, California 95616, U.S.A.

Improving Nutritional Value

Clifford W. Hesseltine, Northern Regional Research Laboratory, 1815 N. University, Peoria, Illinois 61604, U.S.A.

Keith H. Steinkraus, Department of Food Science and Technology, New York State Agricultural Experiment Station, Geneva, New York 14456, U.S.A.

Production of Meat-Like Flavors

Hideo Ebine, National Food Research Institute, Ministry of Agriculture and Forestry, Yatabe-machi, Ibaraki-ken, 300-31, Japan.

Clifford W. Hesseltine, Northern Regional Research Laboratory, 1815 N. University, Peoria, Illinois 61604, U.S.A.

Nakano Masahiro, Meiji Daigaku, Ikuta Kosha, Ikuta 5158, Kawasakishi, Kanagawa-ken 214, Japan.

Shinshu Miso Research Institute, Minamiagata-machi 1014, Nagano-shi 380, Japan.

William Shurtleff, New-Age Foods Study Center, 790 Los Palos Manor, Lafayette, California 94549, U.S.A.

Tamotsu Yokotsuka, Kikkoman Shoyu Co., Ltd., 339 Noda, Noda-shi, Chiba-ken, Japan.

Koji Method of Producing Enzymes

Hideo Ebine, National Food Research Institute, Ministry of Agriculture and Forestry, Yatabemachi, Ibaraki-ken, 300-31, Tokyo, Japan.

Clifford W. Hesseltine, Northern Regional Research Laboratory, 1815 N. University, Peoria, Illinois 61604, U.S.A.

William Shurtleff, New-Age Foods Study Center, 790 Los Palmos Manor, Lafayette, California 94549, U.S.A.

Tamotsu Yokotsuka, Kikkoman Shoyu Co. Ltd., 339 Noda, Noda-shi, Chiba-ken, Japan.

Indonesian Tempeh

Clifford W. Hesseltine, Northern Regional Research Laboratory, 1815 N. University, Peoria, Illinois 61604, U.S.A.

Keith H. Steinkraus, Department of Food Science and Technology, New York State Agricultural Experiment Station, Geneva, New York 14456, U.S.A.

A. G. van Veen, Division of Nutritional Sciences, Cornell University, Savage Hall, Ithaca, New York 14853, U.S.A.

Single-Cell Protein Production

Allen I. Laskin, EXXON Research and Engineering Company, P.O. Box 45, Linden, New Jersey 07036, U.S.A.

John H. Litchfield, Battelle Columbus Laboratories, 505 King Avenue, Columbus, Ohio 43201, U.S.A.

Jacques C. Senez, Centre National de la Recherche Scientifique, Laboratoire de Chemie Bactérienne, 31 Chemin Joseph-Aiguier 13274, Marseille 2, France.

Steven R. Tannenbaum, Massachusetts Institute of Technology, Cambridge, Massachusetts 02139, U.S.A.

Chapter 3

Soil Microbes in Plant Health and Nutrition

The parts of a plant above the ground can be compared to the tip of an iceberg, in that the portion under the surface—the root system—is so extensive. The root system is also very active metabolically and provides a continuous source of food for soil microorganisms in the form of secretions of organic compounds and sloughed-off dead cells and cell debris. Since the zone where roots and soils meet is a special environment, it has been named the rhizosphere (root zone). This zone comprises several poorly defined, heterogeneous regions in which microorganisms are particularly active (Table 3.1).

Although activity in the rhizosphere is of great importance to the plant, it affects only a small fraction (about 5 percent) of the root surface. Some microorganisms are loosely associated with roots, but others develop on the root surface and many can invade root tissue, with effects that can be beneficial or harmful. Certain soil-inhabiting microorganisms produce diseases of great significance to agriculture and forestry. Others are beneficial—they partly inhibit the growth of disease organisms or kill them. The vast majority of the pathogens that infect roots are fungi, and they are exceptionally difficult to control or eradicate.

In some cases, the invasion of roots by microorganisms is desirable. This is true for the root-nodule bacteria of the genus *Rhizobium* that fix nitrogen, as well as for mycorrhizal fungi, which assist roots in accumulating phosphate and other essential minerals. Nitrogen fixation is the subject of Chapter 4 in this report; mycorrhizal fungi are discussed in this chapter. However, it should be emphasized that rhizosphere microorganisms can affect plant welfare in a number of ways that are not yet well understood or easily controlled. The processes of nutrient cycling, growth stimulation or inhibition, and diseases are of great significance, but they are very complex population effects rather than the result of simple interactions between roots and known microorganisms. The rhizosphere is also influenced by external factors such as soil moisture and even the intensity of light reaching the plant. No single microorganism may be essential to the process, but the combined effect of the rhizosphere population can be profound.

TABLE 3.1 Comparison of the Numbers of Various Groups of Organisms in the Rhizosphere of Spring Wheat and in Control Soil

Organisms	Numbers per g in Rhizosphere Soil (× 10^{-6})	Numbers per g in Control Soil (× 10^{-6})	Approximate Rhizosphere: Soil Ratio
Bacteria	1,200	53	23 : 1
Actinomycetes	46	7	7 : 1
Fungi	12	0.1	120 : 1
Protozoa	0.0024	0.001	2 : 1
Algae	0.005	0.027	0.2 : 1
Bacterial Groups			
Ammonifiers	500	0.04	12,500 : 1
Gas-producing anaerobes	0.39	0.03	13 : 1
Anaerobes	12	6	2 : 1
Denitrifiers	126	0.1	1,260 : 1
Aerobic cellulose decomposers	0.7	0.1	7 : 1
Anaerobic cellulose decomposers	0.009	0.003	3 : 1
Spore formers	0.930	0.575	2 : 1
"Radiobacter" types	17	0.01	1,700 : 1
Azotobacter	<0.001	<0.001	?

Adapted from: T. R. G. Gray, and S. T. Williams. 1975. *Soil Microorganisms*, New York: Longman, p. 144.

The sum of the various interrelationships of rhizosphere microorganisms and roots can benefit plant growth by influencing the availability of essential nutrients, by producing plant growth regulators, and by suppressing root pathogens.

Mineral Cycling by Soil Microorganisms

By decomposing plant and animal residues, soil microorganisms release carbon, nitrogen, sulfur, phosphorus, and trace elements from organic materials in forms that can be absorbed by plants.

This process, known as mineralization, is the primary source of atmospheric carbon dioxide. Without mineralization of organic carbon, the carbon dioxide content of the air, which is essential for plant photosynthesis, would be progressively reduced and plant production would ultimately cease. Maintenance of the carbon cycle, therefore, is one of the most important biological processes on earth.

Microbial activities similar to those responsible for the carbon cycle also transform soil nitrogen, sulfur, and phosphorus, and to a lesser extent are instrumental in the conversion of other elements. Although particular attention has been directed to microbial transformations of nitrogen, plants also have a nutritional need for sulfur. The microbial transformations of nitrogen and sulfur are much alike because both elements can be oxidized and reduced. Sulfur reduction is necessary for the synthesis of sulfur-containing amino acids. Under anaerobic conditions, sulfur reduction may produce hydrogen sulfide, which can be harmful to plants. It accumulates in very wet soils, like rice paddies, and can cause straighthead disease of rice and other physiological plant disorders. At low concentrations, however, it can supply the sulfur requirements of some plants.

A recent report indicates that oxidation of hydrogen sulfide by a bacterium in the genus *Beggiatoa* detoxified flooded rice soils. *Beggiatoa* species may be significant in coastal marshes and estuaries as well as in rice paddies, and their capacity to influence plant growth favorably deserves further study. The oxidation of sulfur by bacteria of the genus *Thiobacillus* is also of potential significance in agriculture. The product of this transformation is sulfuric acid, which can dissolve minerals that otherwise would not be available for plant growth. Farmers who add elemental sulfur to rock phosphate find that phosphorus is liberated more rapidly and in greater amounts than if the sulfur is omitted. The explanation for this is that *Thiobacillus* species oxidize the sulfur to sulfuric acid, which liberates the phosphorus from the insoluble rock phosphate.

Microorganisms are also able to promote phosphorus solubilization by the production of chelators, which form complexes with metal ions and increase their solubility. Acid and chelate production can be easily demonstrated under laboratory conditions, but little is known of the phosphate-dissolving effectiveness of different types of microorganisms under natural soil conditions. Solvent action by microorganisms is not restricted to a few species; it is characteristic of many members of the rhizosphere population and can be accomplished in part by plant roots as well. The microbial transformation of nutrient elements other than those cited above is in some instances similar to, and in other instances quite different from, the process just discussed. Unfortunately, much more is still to be learned about this subject.

Microorganisms require many of the same nutrient elements that are essential to plants for their growth. When nitrogen, sulfur, or phosphorus is in short supply, the rhizosphere population will compete with roots for nourishment. Because of their abundance, small size, and relatively large surface area, and because they surround the absorbing part of the root, microbes will absorb nutrients at the expense of the plant. Ultimately, plants will display signs of nutrient deficiency and crop yields may decrease.

Barber and Martin (1976) have recently found that for barley, 10–20 percent of the photosynthate may be released from roots in nonsterile soil. Less was released in sterile soil. The rhizosphere may exact a price in terms of energy given up to the soil by the plant.

There has been speculation that in the rhizosphere oxygen consumption occurs more rapidly than diffusion, so that anaerobic sites may form in places in the root; such reduced conditions could be important in making ferrous ions from ferric, for instance, which increases iron solubility. Wheat roots have high populations of denitrifying bacteria, so oxygen-free conditions must exist in their presence.

Microorganisms are prolific producers of vitamins, amino acids, hormones, and other growth-regulating substances. Many bacteria and fungi isolated from soil are able to synthesize compounds that provoke a growth response in plant tissue. Some produce indoleacetic acid or gibberellins, which are hormones that control plant growth, while others produce vitamins. Many may also produce unidentified growth factors. Rhizosphere microorganisms are variously credited with promoting increased rates of seed germination, root elongation, root-hair development, nutrient uptake, and plant growth.

Inoculation

Root uptake of organic compounds has received more attention in the U.S.S.R. than elsewhere, and Russian investigators claim rhizosphere microorganisms influence the quality as well as quantity of tissue produced by plants. Although there are no experimental results that convincingly establish that growth-promoting substances of microbial origin occur in the rhizosphere, speculation persists that such compounds are synthesized in the vicinity of roots and affect crop yields.

Various bacterial fertilizers have been marketed at different times, but commercial preparations known as azotobacterin and phosphobacterin have received the most attention. Azotobacterin is composed of cells of *Azotobacter chroococcum*, a bacterium able, under some conditions, to fix atmospheric nitrogen. Phosphobacterin contains the bacterium *Bacillus megaterium* var. *phosphaticum*, which mineralizes organic phosphorus compounds. Russian scientists think that growth of these bacteria in soil will supply plants with nitrogen and phosphorus, but this has not been proved. Treated plants are favorably affected, but growth is not increased by more than 10 percent. Moreover, the effect is said to be due not to nitrogen fixation or phosphorus solubilization, but to plant hormones.

Since the benefits are minimal and depend on conditions difficult to control in the field, and the results are unpredictable, bacterial fertilizers are not recommended for general use.

Although the potential benefits of inoculating bacteria have not yet been fully explored, it is questionable how much additional exploration may be warranted. Present evidence is insufficient to justify the use of inoculants, other than rhizobia for legumes, to increase crop yields, improve plant quality, or control disease. A beneficial effect is even less likely when the microorganism used as inoculum is a normal inhabitant of soil. The British soil microbiologist S. D. Garrett (1956) has described this problem as follows:

> [Such] attempts to boost the population of an antagonistic organism by inoculation alone have been doomed to failure from their inception, because

they are in flagrant contradiction to the ecological axiom that population is a reflection of the habitat, and that any change due to plant introduction without change of the habitat must be a transient one.

Mycorrhizal Fungi

Most plants, both wild and cultivated, have roots infected with fungi that increase nutrient and water uptake and may also protect the root from certain diseases. These infected roots are called mycorrhizae. Although the mycorrhizal fungi probably increase uptake of all the essential elements, they are usually most important in improving phosphorus nutrition. Phosphate is generally present in the soil in low concentrations and it is also highly immobile. Strands of fungal hyphae grow out from mycorrhizae and greatly increase the volume of soil from which phosphorus is obtained. So mycorrhizal plants, in general, can grow and thrive in soils much lower in phosphate and other essential nutrients than a comparable nonmycorrhizal plant. Many plants are so dependent on mycorrhizal fungi for nutrient uptake that they may starve if these fungi are absent. There are a number of types of mycorrhizae. The two that occur on the most economically important crops, the endomycorrhizae and the ectomycorrhizae, are discussed below.

Endomycorrhizae of Crop Plants and Forest Trees

Endomycorrhizae of the vesicular-arbuscular (VA) type occur on nearly all crop plants (plants in a few families such as the *Cruciferae* [cabbage, mustard, etc.] and *Chenopodiaceae* [beets, spinach, etc.] may be nonmycorrhizal). They also occur on many trees in temperate regions and on the majority of tree species native to the subtropics and tropics. VA mycorrhizal fungi are present in almost all soils and they are not host-specific. Thus, the same fungus producing VA mycorrhizae on trees will form mycorrhizae on plants after land is cleared and planted to agricultural crops. The mycorrhizal condition is normal for most plants, and absence or scarcity of mycorrhizal fungi can greatly limit plant growth (Figure 3.1). Introduction of mycorrhizal fungi to soil environments lacking or with low populations of such organisms, such as biocide-treated soils, can enhance plant growth. VA mycorrhizae are particularly important for many legumes in that they stimulate nodulation by *Rhizobium*, thereby increasing nitrogen fixation. Improved phosphorus nutrition of the mycorrhizal legume is responsible for increased nodulation.

VA mycorrhizal fungi survive in soil as resting spores. They obtain their food from the plant roots and they are unable to grow independently in soil. It is unlikely that they obtain much, if any, organic nutrient from soil.

VA fungi have not been grown in pure culture, which presents an obstacle to artificial inoculation. However, these fungi produce the largest spores of

FIGURE 3.1 Endomycorrhizal (left) and nonmycorrhizal (right) peanut plants (groundnuts). (Photograph courtesy of J. W. Gerdemann)

any known fungi, some being 0.5 mm or more in diameter. The spores can be easily extracted from the soil with sieves and then propagated on the roots of living plants. The infected roots can also be used for inoculation. Heavily infected palm roots collected in the wild have been used as a source of inoculum. If field-collected inoculum is used, it is important that it be free of dangerous pathogens.

There are situations in which inoculation with VA fungi is highly beneficial. If soil is treated with steam or fumigants to kill pathogens, VA fungi are also killed, and considerable time is required for them to become reestablished naturally. The nutrient deficiencies and associated stunting that often result may be prevented by inoculating the soil with VA fungi rather than by applying excessive rates of fertilizer.

The greatest opportunity for the use of VA fungi is in soils low in available phosphorus, which includes many untreated soils in tropical regions. There is evidence as well that many of these soils also contain less than the optimum number of spores of VA fungi. In such soils inoculation may enable the use of inexpensive rock phosphate fertilizer instead of the more expensive super and triple phosphates.

The major obstacle to greater use of VA fungi is the difficulty in obtaining inoculum. However, there are several commercial companies interested in producing pure inoculum, and it may become available in the near future. We are now at the stage where different species of VA fungi should be tested on crop plants and forest trees and their effects on growth compared in the field in untreated soils and in soils that have been sterilized.

Ectomycorrhizae of Forest Trees

Ectomycorrhiza is the second most common type of mycorrhizae. It occurs on roots of pine, spruce, fir, larch, hemlock, willow, poplar, hickory, pecan, oak, birch, beech, and eucalyptus (Figure 3.2). The fungi that form ectomycorrhizae produce mushrooms and puffballs as their reproductive structures (fruit bodies). In North America there are more than 2,100 species of ectomycorrhizal fungi. The fungi are spread in nature by millions of microscopic spores, finer than dust, which are released from fruiting bodies and moved great distances by winds.

Many forest trees, such as pines, cannot grow beyond the first year without an appreciable number of ectomycorrhizae. Ectomycorrhizae benefit trees by: increasing nutrient and water absorption from soil; increasing the tolerance of the tree to drought and extremes of soil conditions (acid levels, toxins, etc.); increasing the length of the feeder root system; and protecting the fine feeder roots from certain harmful soil fungi.

Ectomycorrhizal fungi cannot grow and reproduce unless they are in association with the roots of a tree host. These fungi obtain all their essential

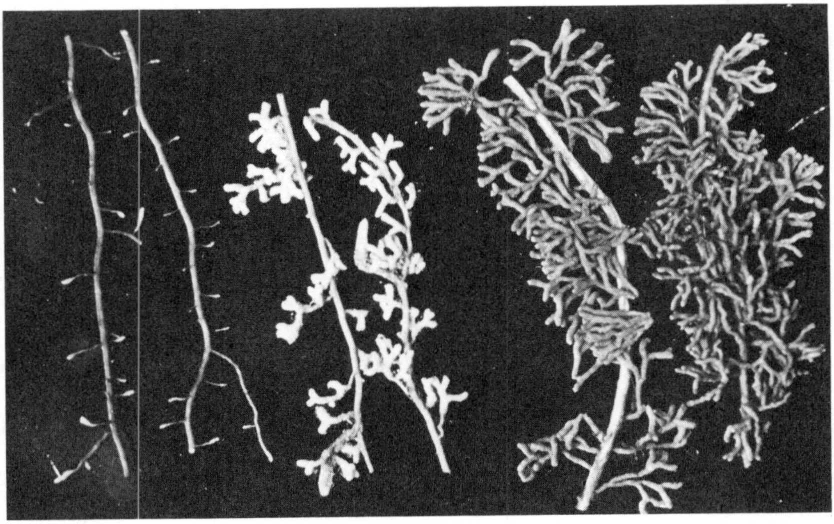

FIGURE 3.2 Examples of pine ectomycorrhizae. Each different ectomycorrhiza is formed by a different species of fungus. Each main root is approximately 3 cm long. (Photograph courtesy of J. W. Gerdemann)

sugars, vitamins, amino acids, and other foods from their hosts. It is unlikely that these fungi as a group are directly involved in any significant decomposition of forest litter.

Certain forest trees, then, must have ectomycorrhizae to survive and grow, and the ectomycorrhizal fungi need their tree hosts to exist. This means that the introduction of tree species as exotics into regions where the appropriate mycorrhizal fungi are absent must be accompanied by the introduction of their natural ectomycorrhizal fungi. In the past, this introduction has been accomplished mainly by using soil collected from under healthy trees with ectomycorrhizae, which is mixed into the upper layer of soil in nurseries. The seedlings planted in this soil usually form abundant ectomycorrhizae in one growing season, and they are then transplanted to the field. Unfortunately, this method is not without risk, since pathogens can be present in the introduced soils and cause serious damage to the trees. The logistics of transporting large volumes of soil great distances is an added problem. By far the most biologically sound method of correcting an ectomycorrhizal deficiency is by the use of pure cultures of selected species of ectomycorrhizal fungi.

In recent years, techniques have been devised to inoculate soil with pure vegetative and spore cultures of *Pisolithus tinctorius* in the United States and to introduce spore cultures of *Rhizopogon luteolus* onto pine seed and into soil in Australia. These two puffball-producing fungi form ectomycorrhizae on many commercially important forest trees. Currently, research is being done in the United States on the use of a commercially produced vegetative inoculum of *P. tinctorius,* which should be available at economical prices on the world market in the next few years. Thus far, *P. tinctorius* appears to enhance growth more than other ectomycorrhizal fungi, and it can be used to tailor-make seedlings to improve the performance of trees even in areas where other ectomycorrhizal fungi are present.

Pines with *Pisolithus* ectomycorrhizae formed in nurseries and planted in forest sites in the southern United States have not only survived, but have grown to twice the heights of comparable pines with naturally occurring ectomycorrhizae. On sites where it is difficult to establish pines, such as those created by strip-mining for coal, the only trees capable of growing are often those that have been inoculated with *Pisolithus.* The selection and use of specific ectomycorrhizal fungi may well determine whether a productive forest becomes established.

The most obvious research need on ectomycorrhizae in developing countries is to determine whether appropriate ectomycorrhizal fungi are present prior to the establishment of forests of introduced species. If such forests are already established or are accessible, then fruit bodies of ectomycorrhizal fungi can be collected and the spores harvested. The puffball fungi usually produce an abundance of easily extractable spores. For example, one fruit body of *P. tinctorius* may contain 75 grams of spores, and there are more than one billion spores in a gram. Spores of *P. tinctorius* when kept dry and cool have been stored for more than 4 years without losing their ability to form ectomycorrhizae. The spores can be mixed into nursery soil and the seed of the desired tree species planted. Moderate levels of fertility and at least 2–4 percent organic matter should be maintained in the soil. Usually, the seedlings will have adequate ectomycorrhizae in 6–7 months after seed germination.

The production of vegetative cultures of ectomycorrhizal fungi requires aseptic culture technique. This means that the substrate on which the fungi are grown must be sterilized, usually by autoclaving, and maintained free of other microbial contamination for at least several weeks, or until the fungus has produced sufficient growth to overcome contamination. After it has been leached with water, this inoculum can then be added to soil.

Some species of ectomycorrhizal fungi are more beneficial to tree growth and development under different soil conditions than others. It is important to select and test different species to determine which are best suited to specific locations.

Biological Control of Soil-Borne Pathogens

Microorganisms that cause root diseases are sometimes suppressed by other microorganisms in the soil. In many instances disease-causing organisms may be present, but because of naturally occurring biological control, little or no disease results. The prevalence of pathogens may be reduced by crop rotation using a non-host crop, which often starves the pathogen and prevents it from reproducing.

It is also possible to increase the level of organisms that are antagonistic to soil-borne plant pathogens. This is generally done not by adding antagonistic microorganisms directly to the soil, but by the use of various organic amendments such as manure or plant residues. These amendments provide a source of food for soil-borne microorganisms that can inhibit the development of plant pathogens. Research is needed in order to exploit more fully the use of various forms of organic matter to enhance this biological control of soil-borne pathogens.

There have been many attempts to control pathogens in soil by the addition of specific microorganisms. In general, these attempts have failed to increase the level of naturally occurring biological control. For example, soil contains species of fungi that trap and feed on plant parasitic nematodes (Figure 3.3). However, application of additional nematode-trapping fungi failed to protect plants under field conditions. It is likely that unless the soil is altered in some way, it naturally contains the maximum number of nematode-trapping fungi that it can support. There is, however, hope that we may

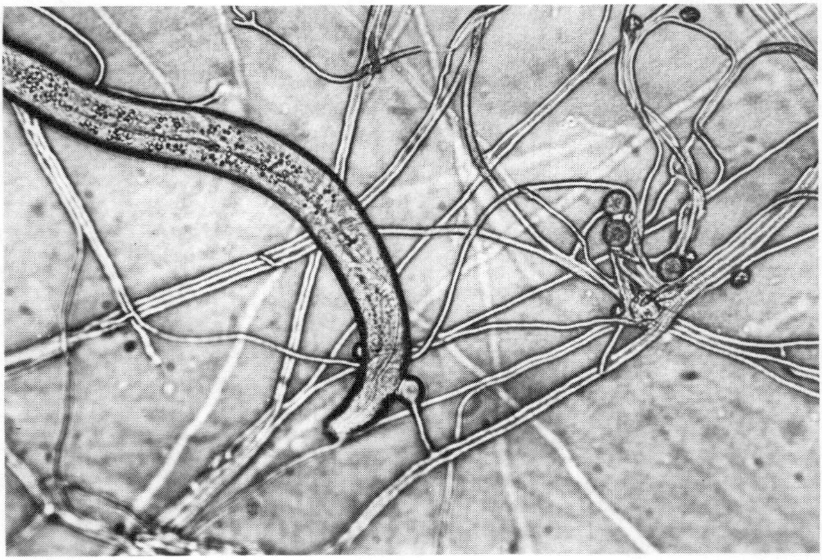

FIGURE 3.3 A nematode-trapping fungus, *Dactylella drechsleri*, which captures prey on adhesive knobs. (Photograph courtesy of D. Pramer)

be on the verge of a major advance in controlling soil-borne pathogens by adding specific microorganisms or by altering the rhizosphere environment. There are a few examples where disease has been reduced by applying a hypovirulent strain or a mutant strain of a pathogen that is incapable of producing disease. Such strains may prevent development of the pathogenic strains. A nonpathogenic strain of the crown gall bacterium will thus protect plants from attack by a pathogenic strain.

In soil, most pathogenic fungi must pass through the rhizosphere or must live within this zone. Their success in colonizing or infecting plant roots depends upon their ability to compete with other rhizosphere microorganisms. The chemical and microbiological environment in this zone may be altered slightly, but significantly, to effect changes in the inoculum potential of pathogens, either by selections of plant genotypes that produce such changes or by careful regulation of nitrification in soil.

Research on biological control should be highly encouraged, for it could provide an alternative means of disease control to the use of expensive and often dangerous pesticides.

References and Suggested Readings

Alexander, M. 1977. *Introduction to soil microbiology*. 2nd Edition. New York: John Wiley and Sons.

____ . 1974. *Microbial ecology*. New York: John Wiley and Sons.

Baker, K. F., and Cook, R. J. 1974. *Biological control of plant pathogens*. San Francisco: W. H. Freeman and Company.

Barber, D. A., and Martin, J. K. 1976. The release of organic substances by cereal roots into soil. *The New Phytologist* 76:69-80.

Barber, D. S. 1968. Microorganisms and the inorganic nutrition of higher plants. *Annual Review of Plant Physiology* 19:71-88.

Barron, G. L. 1977. *The nematode-destroying fungi*. Topics in mycobiology No. 1. Guelph, Ontario: Canadian Biological Publications Ltd.

Brown, M. E. 1974. Seed and root bacterization. *Annual Review of Phytopathology* 12:181-197.

____ ; Hornby, D.; and Pearson, V. 1973. Microbial populations and nitrogen in soil growing consecutive cereal crops infected with take-all. *Journal of Soil Science* 24(3):296-310.

Carson, E. W. 1974. *The plant root and its environment*. Charlottesville: The University Press of Virginia.

Cook, R. J. 1977. Management of the associated microbiota. In *Plant disease: an advanced treatise in how disease is managed.* J. G. Horsfall and E. B. Cowling, eds., pp. 145-166. New York: Academic Press.

____ . 1976. Interaction of soil-borne plant pathogens and other microorganisms: an introduction. *Soil Biology and Biochemistry* 8:267.

Doetsch, R. N., and Cook, T. M. 1976. *Introduction to bacteria and their ecobiology.* Baltimore, Maryland: University Park Press.

Garrett, S. D. 1956. *Biology of root-infecting fungi.* p. 11. New York: Cambridge University Press.

____ . 1970. *Pathogenic root-infecting fungi.* Cambridge, England: Cambridge University Press.

Gerdemann, J. W. 1975. Vesicular-arbuscular mycorrhizae. In *The development and function of roots.* J. G. Torrey and D. T. Clarkson, eds., pp. 575-591. New York: Academic Press.

Gray, T. P., and D. Parkinson. 1968. *The ecology of soil bacteria.* Liverpool: Liverpool University Press.

____ , and Williams, S. T. 1975. *Soil microorganisms.* New York: Longman.

Harley, J. L. 1979. *Proceedings of the soil-root interface symposium*: London: Academic Press.

Henis, Y., and Chet, I. 1975. Microbiological control of plant pathogens. *Advances in Applied Microbiology* 19:85.

Hornby, D. 1978. Microbial antagonisms in the rhizosphere. *Annals of Applied Biology* 89:97-100.

Joshi, M. M., and Hollis, J. P. 1977. Interactions of *Beggiatoa* and rice plants: detoxification of hydrogen sulfide in the rice rhizosphere. *Science* 197:179-180.

Kleinschmidt, G. D., and Gerdemann, J. W. 1972. Stunting of citrus seedlings in fumigated nursery soils related to the absence of endomycorrhizae. *Phytopathology* 62:1447-1453.

Krasilnikov, N. A. 1958. *Soil microorganisms and higher plants.* Moscow: Academy of Sciences USSR. English translation, by Y. Halperin, 1961. Jerusalem: Israel Program for Scientific Translations, Ltd.

Marks, G. C., and Kozlowski, T. T. 1973. *Ectomycorrhizae: their ecology and physiology.* New York: Academic Press.

Marx, D. H. 1977. The role of mycorrhizae in forest production. In *Proceedings of the TAPPI (Technical Association of the Pulp and Paper Industry) Annual Meeting,* February 14-16, 1977, held in Atlanta, Georgia, pp. 151-161. Atlanta: TAPPI.

Mosse, B. 1977. Plant growth responses to vesicular-arbuscular mycorrhiza: X. Responses of stylosanthes and maize to inoculation in unsterile soils. *New Phytologist* 78:277-288.

____ . 1977. The role of mycorrhiza in legume nutrition on marginal soils. In *Exploiting the legume-rhizobium symbiosis in tropical agriculture:* Proceedings of a workshop,

University of Hawaii, August 1976, College of Tropical Agriculture Miscellaneous Publication No. 54, pp. 275-292. Honolulu: University of Hawaii.

Rovira, A. D.; Newman, F. I.; Bowen, H. J.; and Campbell, R. 1974. Quantitative assessment of the rhizoplane microflora by direct microscopy. *Soil Biology and Biochemistry* 6:211-216.

Sanders, F. E.; Mosse, B.; and Tinker, P. B., eds. 1974. *Endomycorrhizas: Proceedings.* Symposium on Endomycorrhiza. July, 1974, University of Leeds. New York: Academic Press.

Shipton, P. J. 1977. Monoculture and soilborne pathogens. *Annual Review of Phytopathology* 15:387-407.

Smith, A. M. 1976. Availability of plant nutrients in reduced microsites in soil. *Annual Review of Phytopathology* 14:53-73.

Tansey, M. R. 1977. Microbial facilitation of plant mineral nutrition. In *Microorganisms and minerals*, E. D. Weinberg, ed., pp. 343-385. New York: Marcel Dekker, Inc.

United Nations Educational, Scientific, and Cultural Organization. 1969. *Soil biology: review on research.* Natural Resources Research, Series No. 9. Paris: United Nations Educational, Scientific, and Cultural Organization. Distributed in the United States by UNIPUB, New York.

Research Contacts

Rhizosphere

Richard Bartha, Rhizosphere Group, Cook College, Rutgers University, New Brunswick, New Jersey 08903, U.S.A.

J. P. Hollis, Department of Plant Pathology, Louisiana State University, Baton Rouge, Louisiana 70803, U.S.A.

A. D. Rovira, Division of Soils, CSIRO, P.O. Box 2, Glen Osmond, Adelaide, South Australia, 5064.

Endomycorrhizae

J. W. Gerdemann, Department of Plant Pathology, University of Illinois, Urbana, Illinois 61801, U.S.A.

John A. Menge, Department of Plant Pathology, University of California, Riverside, California 92521, U.S.A.

Barbara Mosse, Rothamsted Experimental Station, Harpenden, Hertshire AL5 2JQ, England

T. H. Nicolson, Department of Biological Sciences, University of Dundee, Dundee, DD1 4HN, Scotland

Ectomycorrhizae

Donald H. Marx, Director and Chief Plant Pathologist, Institute for Mycorrhizal Research and Development, Forestry Sciences Laboratory, U.S. Department of Agriculture, Carlton Street, Athens, Georgia 30602, U.S.A.

Orson K. Miller, Jr., Department of Biology, Virginia Polytechnic Institute, Blacksburg, Virginia 24061, U.S.A.

James M. Trappe, Forest Service, Forestry Sciences Laboratory, U.S. Department of Agriculture, 3200 Jefferson Way, Corvallis, Oregon 97331, U.S.A.

Biological Control of Soil-Borne Pathogens

R. James Cook, Regional Cereal Disease Research Laboratory, U.S. Department of Agriculture, Washington State University, Pullman, Washington 99163, U.S.A.

Allen Kerr, Department of Plant Pathology, Waite Agricultural Institute, University of Adelaide, Glen Osmond, Adelaide, South Australia, 5064.

Chapter 4

Nitrogen Fixation

Air is four-fifths nitrogen, yet it is the absence of this particular element that most commonly limits food production. Neither man, animals, nor higher plants can use elemental nitrogen; it must first be "fixed," that is, combined with other elements such as hydrogen, carbon, or oxygen before it can be assimilated.

Certain bacteria and algae have the ability to utilize (fix) gaseous nitrogen from the air. Some microorganisms work symbiotically in nodules on the roots of plants, with the plant providing food and energy for the bacteria, which, in turn, fix nitrogen from the air for their host. Other kinds of bacteria and algae work independently and fix nitrogen for their own use, but these are often limited in their activity because of the lack of a dependable energy supply.

Bacteria that fix nitrogen in nodules on the roots of leguminous plants are called rhizobia (Figure 4.1). Other microorganisms that produce nodules on certain nonleguminous plants are classified as *Frankia* spp. and are actinomycetes. Freshly crushed nodules from the same plant species also will induce nodulation on these plants. Recently Callaham *et al.* (1978) have induced nodulation in shrubs of sweet fern (*Comptonia peregrina*) with a pure culture of *Frankia.*

Leguminous plants have been known for centuries to enrich soils, but the reason was not understood until 1886 when two German scientists, Hellriegel and Wilfarth, found that the bacteria in the nodules on the leguminous root brought about nitrogen fixation.

Nitrogen-fixing microorganisms fix an estimated 175 million t of nitrogen annually, or about 70 percent of our total supply. The remainder is produced in chemical fertilizer factories. With the rising world population and the declining supply of fossil fuels required to manufacture chemical nitrogen fertilizer, it may be necessary to rely more on microorganisms to satisfy plant needs for nitrogen. Some of the nitrogen-fixing systems involving microorganisms are described in the following sections.

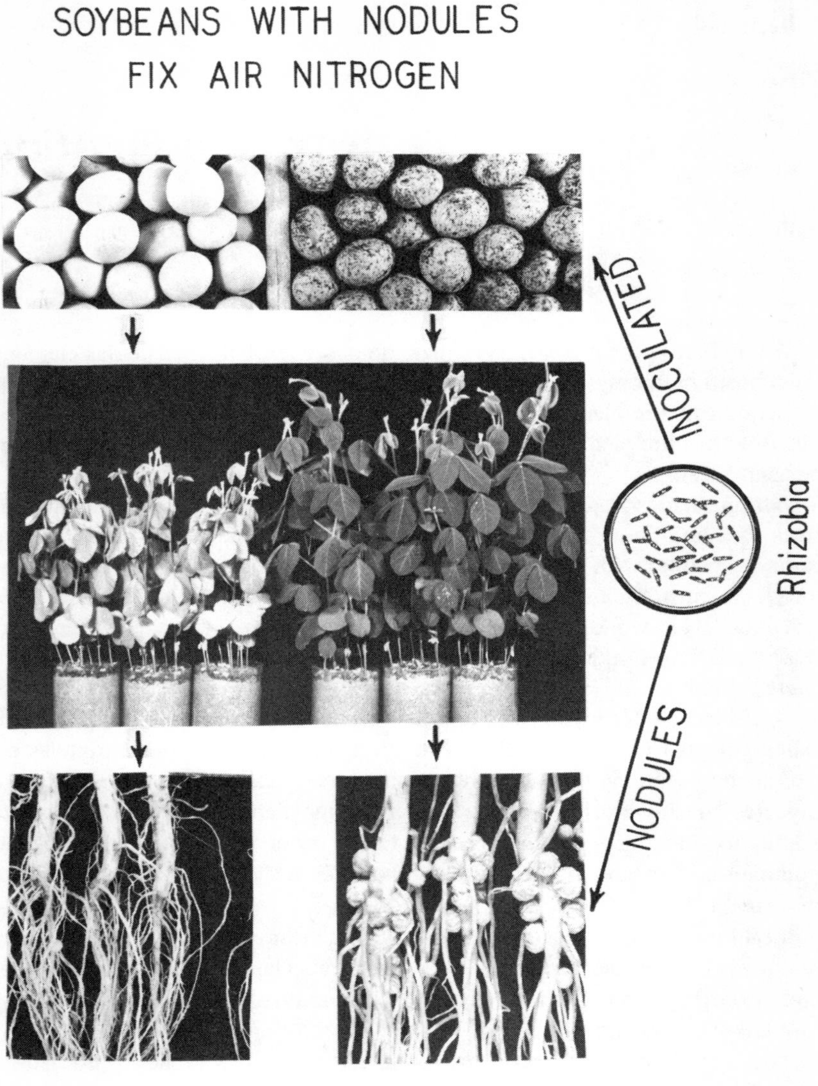

FIGURE 4.1 Rhizobia of the proper kind applied to leguminous seeds before planting can induce nodules or nitrogen factories to form on the roots. These provide the plant with usable nitrogen. The process of applying the rhizobia to seed is called inoculation. (Photograph courtesy of Joe C. Burton)

Symbiotic Systems

Rhizobium-Leguminous Plant Associations

Of all the systems of biological nitrogen fixation, the *Rhizobium*-leguminous plant association has been the most reliable. Legumes in association with nodule bacteria fix at least 35 million t of nitrogen annually valued at several billion U.S. dollars. Yet this beneficial association of nodule bacteria with legumes has only been partially explored. Of the 13,000 known species of legumes, only about 100 are grown commercially. Further, much of the seed is planted without inoculations with the nodule bacteria. Nodulation, if it occurs under these conditions, is by native soil rhizobia, which are often either ineffective or too few in number to bring about effective nitrogen fixation.

Plant Groups and *Rhizobium* Species. Many soils do not contain the proper nodule bacteria to bring about nitrogen fixation and successful growth of legumes. In many cases when the bacteria are present, they are ineffective—that is, they produce nodules that provide little or no nitrogen. Farmers can enhance nitrogen fixation by adding the proper nitrogen-fixing bacteria to leguminous seeds before they are planted. Less than a kilogram of high-quality inoculant, properly applied to legume seeds, can replace more than 100 kilograms of fertilizer nitrogen per hectare.

Certain groups of leguminous plants are nodulated by a single kind of *Rhizobium*. The bacteria that nodulate each of these groups are often considered a species. All plants susceptible to nodulation by a *Rhizobium* species constitute a "cross-inoculation" group. The *Rhizobium* species and their corresponding plant or cross-inoculation groups are given in Table 4.1.

Effective nitrogen-fixing nodules on some common legumes are shown in Figure 4.2. These nodules are usually large and are often concentrated on the primary root. In contrast, ineffective nodules are small, numerous, and scattered over the root system (Figure 4.3).

Rhizobium-Host Interactions. Strains of rhizobia cannot be described as effective or ineffective without specifying the exact species of legume host. Strains of rhizobia that are good nitrogen-fixers in association with one host are often worthless on another. There is voluminous literature on this point, but so many of the leguminous plants cultured in the tropics and subtropics are nodulated by the cowpea rhizobia that special mention is justified. The cowpea cross-inoculation group encompasses numerous genera and species of plants. *Rhizobium* strains effective on a wide spectrum of plants within this group are scarce. Specific inocula containing two or three strains known to be highly effective on the particular legume may be needed to assure good yields.

TABLE 4.1 *Rhizobium* Species and Plants Nodulated

Rhizobium Species	Plants Nodulated
Designated	
R. meliloti	*Medicago sativa* (alfalfa)
	Melilotus sp. (sweet clover)
	Medicago sp. (burr and barrel medic annuals)
	Trigonella foenum graecum (fenugreek)
R. trifolii	*Trifolium* sp. (clovers)
R. leguminosarum	*Pisum sativum* (garden and field peas)
	Vicia faba (broad bean)
	Lens esculenta (lentils)
	Lathyrus sp. (peavine)
R. phaseoli	*Phaseolus vulgaris* (common, field, haricot, kidney, pinto, snap beans, etc.)
	P. coccineus (runner bean, scarlet runner)
R. lupini	*Lupinus* sp. (all lupins)
	Ornithopus sativus (serradella)
R. japonicum	*Glycine max* (soybean)
Undesignated	
Rhizobium sp.	*Vigna unguiculata* (cowpea)
(Cowpea type)	*Arachis hypogaea* (peanut, groundnut)
	Vigna radiata (mung bean)
	Phaseolus lunatus (lima bean)
	P. acutifolius (tepary bean)
	Psophocarpus tetragonolobus (winged bean)
	Sphenostylis sp. (African yam bean)
	Pachyrhizus sp. (jicamus)
	Centrosema sp. (centro)
	Mucuna deeringiana (velvet bean)
	Canavalia ensiformis (jack bean)
	Lablab purpureus (hyacinth bean)
	Phaseolus aconitifolius (moth bean)
	Cyamopsis tetragonoloba (guar)
	Voandzeia subterranea (Bambara groundnut)
	Cajanus cajan (pigeon pea)
	Desmodium sp.
	Cassia sp.
	Lespedeza sp.
	Indigofera sp.
	Crotalaria sp.
	Pueraria sp.
Rhizobium sp.	*Cicer arietinum* (chick-pea, garbanzo)
	Coronilla varia (crownvetch)
	Onobrychis vicisefolia (sainfoin)
	Leucaena leucocephala (ipil-ipil)
	Petalostemum sp. (prairie clover)
	Albizzia julibrissin
	Lotus sp. (trefoils)
	Anthyllis vulneraria (kidney vetch)
	Sesbania sp.

Sources: R. E. Buchanan, and N. E. Gibbons, eds. 1974. *Bergey's Manual of Determinative Bacteriology*. 8th edition. Baltimore: The Williams and Wilkins Co. E. B. Fred; I. L. Baldwin; and E. McCoy. 1932. *Root-Nodule Bacteria and Leguminous Plants.* Madison: University of Wisconsin Press.

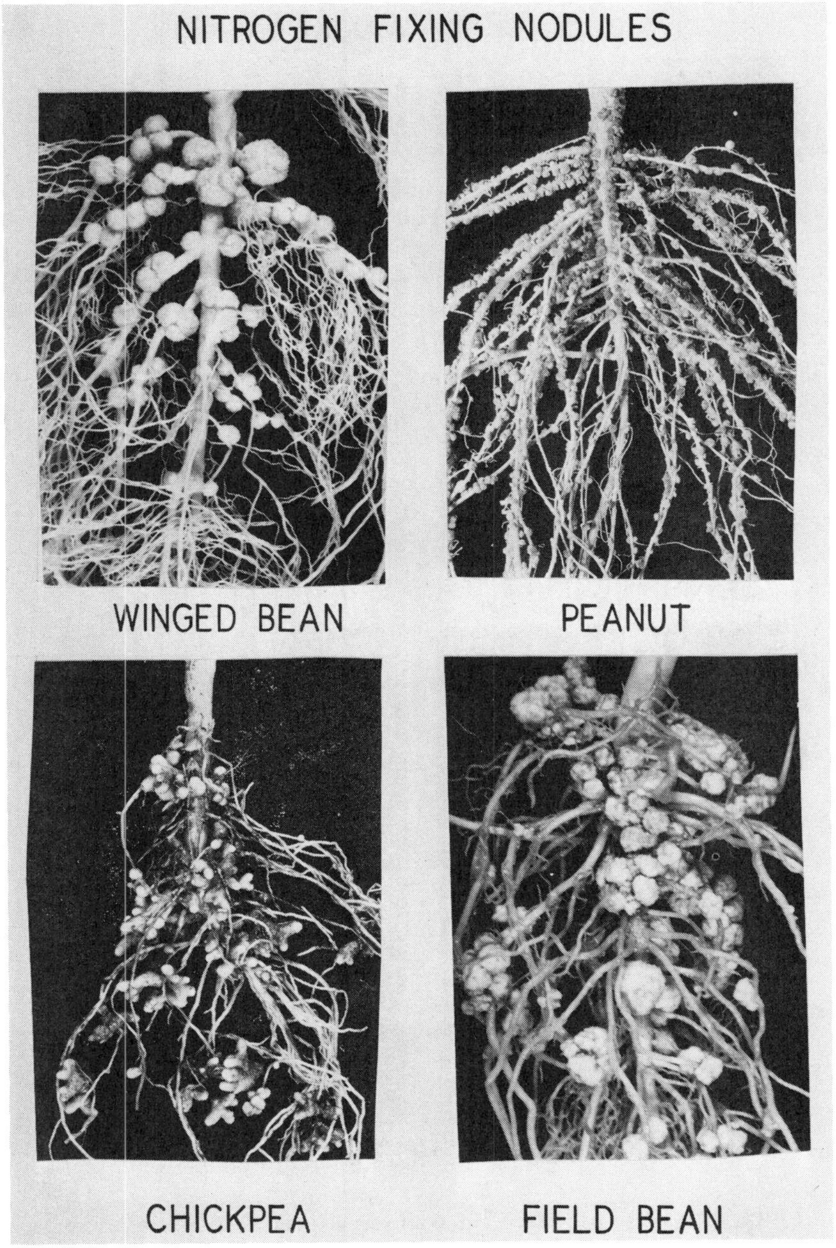

FIGURE 4.2 Nodules or nitrogen factories on the roots of important food legumes: winged bean, *Psophocarpus tetragonolobus*; peanut, *Arachis hypogaea*; chickpea, *Cicer arietinum*; and field bean, *Phaseolus vulgaris*. (Photographs courtesy of Joe C. Burton)

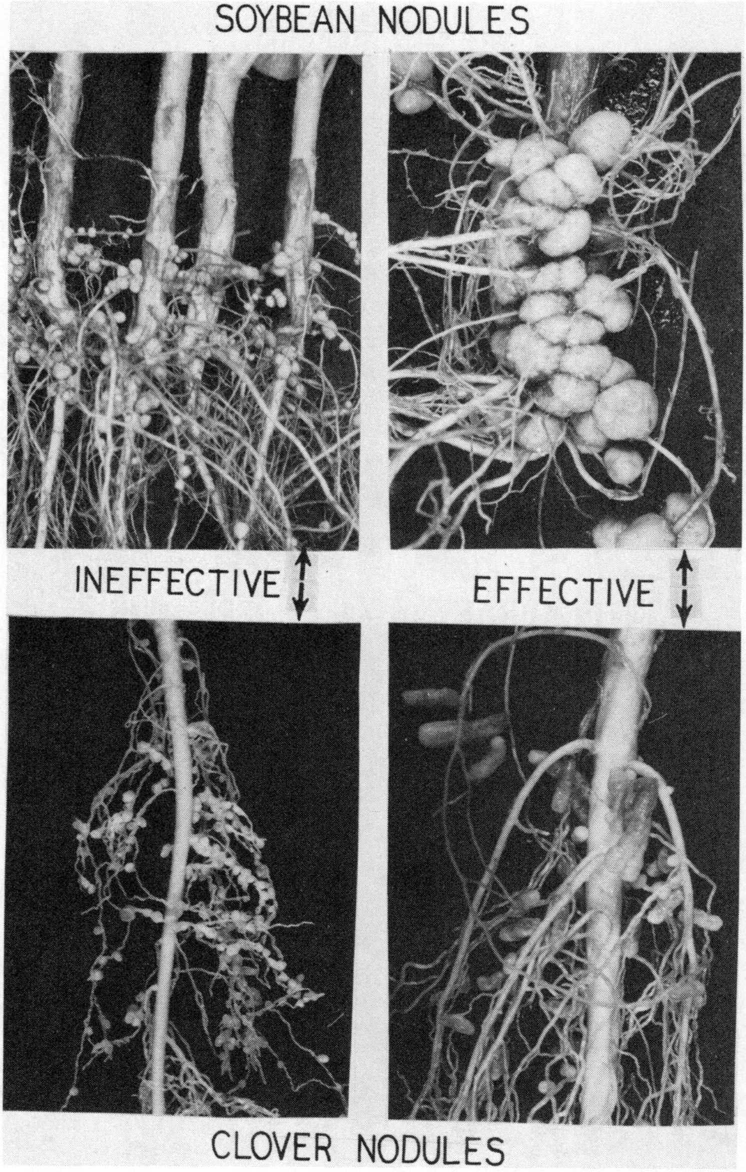

FIGURE 4.3 Rhizobia vary in their nitrogen-fixing abilities. Some are ineffective; they produce nodules and use food that the plant provides, but fix little or no nitrogen. Others are effective; these produce large nodules and fix appreciable amounts of nitrogen. (Photographs courtesy of Joe C. Burton)

Nitrogen Fixed by Leguminous Plants. The amount of nitrogen fixed in *Rhizobium*-leguminous plant associations varies with both the bacteria and legumes as well as environmental factors. Forage legumes usually fix more nitrogen than do grain legumes because carbohydrate requirements resulting from seed development are small, whereas with grain legumes the developing large seeds impose a large demand on the carbohydrate supply. The amounts of nitrogen fixed are uncertain, because of the methods of measurement, but relative quantities as related to host species give valid comparisons (Table 4.2).

Rhizobium species in association with leguminous vegetables, in addition to increasing plant protein, make the plants richer in vitamins and mineral content by assisting general growth. Healthy, well-fed plants are more palatable and nutritious than plants suffering from lack of nitrogen.

Limitations

A major limitation to culture of leguminous crops is the lack of viable, effective inocula for many of the legumes. Leguminous crops and soil conditions vary with each location. *Rhizobium* strains should be selected for particular legumes and soil and climatic conditions. However, *Rhizobium* inoculants are highly perishable and often lose viability before reaching their destination.

TABLE 4.2 Nitrogen Fixed by Various *Rhizobium*-Legume Associations

Plant	Approximate Ranges of Nitrogen Fixed (kg/ha/yr)
Alfalfa, *Medicago sativa*	100–300
Sweet Clover, *Melilotus* sp.	125
Clover, *Trifolium* sp.	100–150
Cowpea, *Vigna unguiculata*	85
Faba bean, *Vicia faba*	240–325
Lentils, *Lens* sp.	100
Lupines, *Lupinus* sp.	150–200
Peanuts, *Arachis hypogaea*	50
Soybeans, *Glycine max*	60–80
Mung bean, *Vigna radiata*	55
Velvet bean, *Mucuna pruriens*	115
Pasture legumes, *Desmodium* sp., *Lespedeza* sp.	100–400

Sources: Adapted from R. C. Burns, and R. W. F. Hardy, 1975. *Nitrogen fixation in bacteria and higher plants.* Berlin: Springer-Verlag; and W. S. Silver, and R. W. F. Hardy, 1976. *Biological nitrogen fixation in forage and livestock systems.* American Society of Agronomy Special Publication No. 28. pp. 1-34.

A limitation to the legume symbiosis system, particularly with the grain legumes, is the relatively short season of active nitrogen fixation in the nodule, especially with plants such as field beans (*Phaseolus vulgaris*). Soybeans fix nitrogen a little longer. Nitrogen fixation with the soybean could be doubled if the period of active nitrogen fixation in the nodule could be increased by as little as 10 days (Hardy and Havelka, 1970).

A limitation to the symbiotic system in leguminous plants is the inadequacy of inoculants and inoculation methods to ensure a greater proportion of nodules from the applied rhizobia when seeds are planted in soils infested with highly infective rhizobia of poor nitrogen-fixing properties.

Research Needs

- *Rhizobium* strains should be selected for the specific legume crops being grown in each country. Strains should be selected for their nitrogen-fixing ability and competitiveness under the prevailing soil and climatic conditions.
- Small-scale inoculant production methods should be studied. Effective use of leguminous plants will depend both on effective *Rhizobium* strains and a dependable delivery system that will help ensure a high production of nodules by the inoculum rhizobia.
- In devising delivery systems, consideration should be given to overcoming an aggressive native population of both infective rhizobia and other microorganisms. The latter group of plant pathogens—as well as insects—may necessitate separate application of inoculants and protective chemicals to the seeds.
- New genera and species of legumes should be studied. Leguminous plants such as the winged bean, *Psophocarpus tetragonolobus*, the climbing lima, *Phaseolus lunatus*, the yam bean, *Sphenostylis stenocarpa*, and the hyacinth bean, *Lablab purpureus*, all of which are climbers, can be very productive under humid tropical conditions. Further, they will fix nitrogen for months, providing the pods are gathered regularly and not allowed to mature on the vine.

Frankia-Nonleguminous Plant Associations

Numerous nonleguminous trees and woody shrubs form root nodules and fix atmospheric nitrogen under natural conditions. The organism responsible, found in the nodules (called an endophyte), is an actinomycete (*Frankia* sp.) and infection and nodule initiation have recently been achieved with the cultured organism (Callaham *et al.*, 1978). Crushed nodules from a growing plant of the same species readily induce nodulation in most cases. Now that the endophyte from one nonleguminous nodulating plant has been isolated and cultivated, this can be done with others, thereby greatly facilitating culture of these nodulating nonleguminous plants. The same organism has now been shown to nodulate *Alnus, Myrica*, and *Comptonia.*

Nonleguminous trees and woody shrubs capable of fixing nitrogen comprise 145 species in 15 different genera of 7 plant families. Within such a large group of plants, there is adaptability to a range of diverse soil and climatic conditions. These plants often thrive in nitrogen-deficient eroded areas and on sand dunes, barren slopes, and even arid soils. Certain species are pioneers. They are the first to grow after glaciers have receded. Others make an early appearance after volcanic eruptions and help develop soils from lava.

The hardiness of this group of plants is partially attributable to their ability to fix nitrogen in association with *Frankia* sp. Many species also benefit from association with external or ectotrophic mycorrhizae, and can thus survive minimum phosphorus levels in the soil. The economic impact of the nodulating nonlegume application is mainly in forestry rather than agricultural crops. Trees of the birch family, especially *Alnus* sp., provide wood for timber in many countries; the red alder (*Alnus rubra* Bong) (Figure 4.4) can fix as much as 300 kg of nitrogen per ha per year.

FIGURE 4.4 Nodules on red alder, *Alnus rubra* Bong. Red alder can fix as much as 300 kg nitrogen per ha per year (Table 4.3) when effectively nodulated. (Photograph courtesy of H. J. Evans)

The nitrogen-fixing abilities of these nonleguminous trees and shrubs when they are well nodulated are almost comparable to those of the *Rhizobium*-leguminous plant associations, providing the stands are kept for several years. Silvester (1977) has tabulated reports on nitrogen fixed by various species (Table 4.3).

The unique abilities of these nodulating nonleguminous trees and shrubs to pioneer in new soils, increase fertility, and enrich the soil for growth of economic crops should not be overlooked. Such plants can provide the base for expansion of areas for food production, and they can help to restore soils disrupted by the removal of coal and minerals.

Limitations

The nodulating nonleguminous trees and shrubs are long-term crops, making them unsuitable when annual harvests of farm crops must be made for subsistence. Recently a new plant (*Datisca cannabina*) was discovered, which nodulates like *Alnus*, is not woody, and is propagated by seed. There may be many others of this type.

Most known valuable species grow best in cool or temperate climates or at high altitudes in the tropics. With further study, species well adapted to the lowland tropics may also be discovered.

Seeds are very limited in availability.

TABLE 4.3 Nitrogen Fixed by Various Genera and Species of Nodulating Nonleguminous Trees and Shrubs

Species	Age–Years	Nitrogen kg/ha/yr*
Alnus crispa	0–5	362
	15–20	115
	10–60	40
Alnus glutinosa	0–8	125
	20	56–130
Alnus rubra	2–15	325
Casuarina equisetifolia	0–13	58
Ceanothus sp.	–	60
Coriaria arborea	14–25	129–192
Dryas drummondii	0–25	12
Hippophae rhamnoides	10–15	15
	13–16	179
Myrica gale	3	9

*These are values reported in the literature; fixation rates vary widely with conditions and should be treated as indicative estimates only, not as definitive rates.
Source: W. B. Silvester, 1977. Dinitrogen fixation by plant associations excluding legumes. In *A treatise on dinitrogen fixation*, R. W. F. Hardy and A. H. Gibson, eds. New York: John Wiley and Sons.

Research Needs

- Improved technology is needed for seed collection, production, and handling for the more promising species.
- Dependable laboratory-produced cultures of the endophytes specific to all important species would be very helpful.
- Surveys should be conducted to determine if nitrogen-fixing species of these nodulating nonleguminous plants grow in the lowland tropics.
- Plants should be evaluated for nitrogen-fixing ability as well as for their value as human and animal food.

Azolla-Anabaena Associations

A small floating freshwater fern, *Azolla pinnata* invades lowland rice fields in Indonesia, southern China, Vietnam, and other tropical areas. The upper lobe of the *Azolla* leaflet contains a large leaf cavity inhabited by the blue-green alga, *Anabaena azolla.* The symbiotic nature of the association is evidenced by two findings: 1) the algae in the leaf cavity have 15-20 percent nitrogen-fixing cells (heterocysts) as compared to 5 percent in free-living *Anabaena* species, and 2) the algae grow very poorly when taken from the leaf cavity and placed on an inorganic medium with no combined nitrogen. Growth is obtained in some cultures when the growth medium is supplemented with an organic compound in the form of 0.5 percent sugar (fructose). But in subsequent studies, nitrogenase activity of this isolated algae has been only about half that of algae growing symbiotically in the leaf.

The *Azolla-Anabaena* association is literally a live floating nitrogen factory, using energy from photosynthesis to fix atmospheric nitrogen. Under Indonesian environmental conditions, the *Azolla-Anabaena* association can fix from 100 to 150 kg of nitrogen per ha per year in approximately 40–60 t of biomass. It is important not to let the fern cover the water completely in the rice paddies. Because the rice can be damaged from excessive shading of the paddy water, 50 percent coverage is ideal. Nitrogen fixation occurs at night as well as during the day, but the rate of fixation is lower in the dark. *Azolla pinnata* plants floating on the water surface of irrigated rice paddies are shown in Figure 4.5.

Azolla has been used extensively in Asia as a forage crop for duck and pig feed, but its greatest potential appears to be as a green manure. On a dry-weight basis it contains about 23.8 percent crude protein, 4.4 percent fat, 6.4 percent starch, and 9.5 percent fiber. Vietnam and Thailand have used *Azolla* for years in their system of rice culture. Stocks of *Azolla* are kept during the hot season for multiplication and distribution when cooler weather comes. The stocks are then used to seed other paddies fertilized with ashes, urine, and rotted manure. *Azolla* vegetation can double approximately every 5 days under favorable conditions.

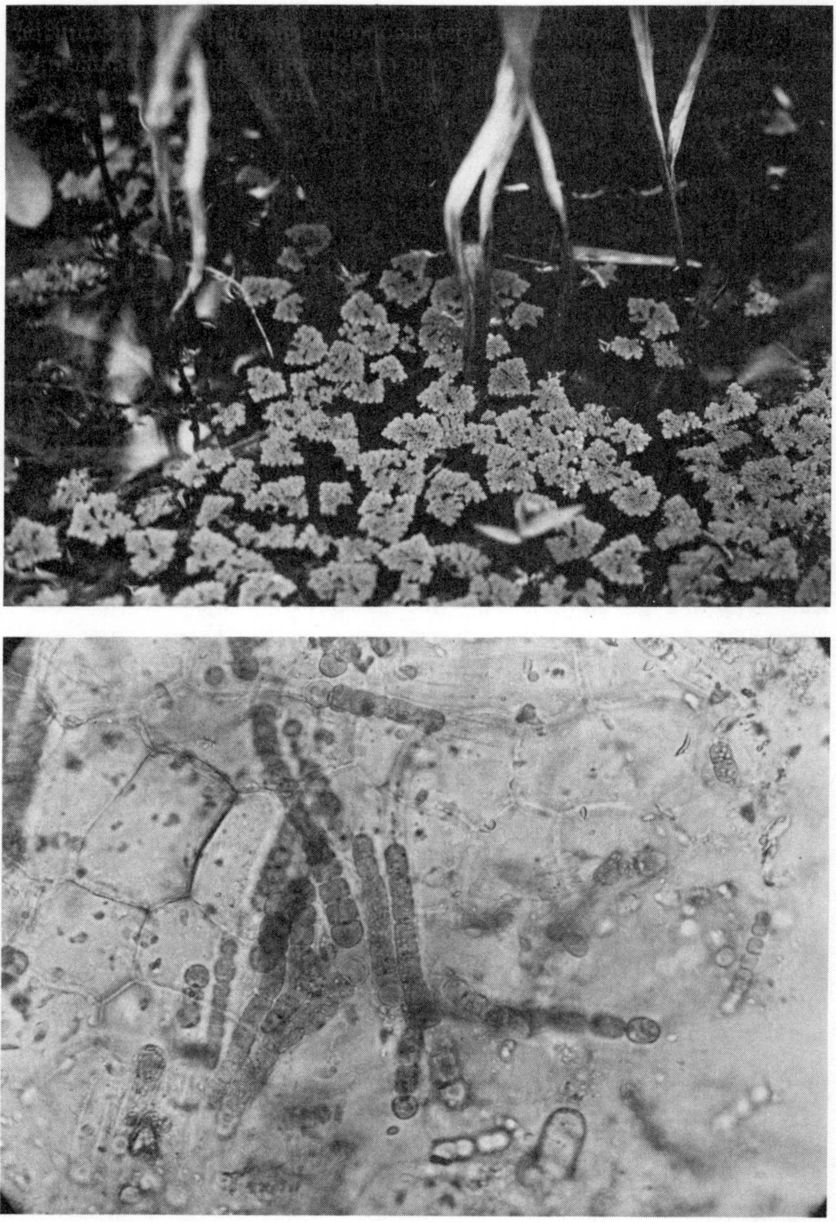

FIGURE 4.5 Small floating nitrogen factories on a flooded rice paddy. The small fern *Azolla pinnata* harbors blue-green algae, *Anabaena azolla*, which fix nitrogen. Together they provide nitrogen for a future crop. (Photograph courtesy of J. H. Becking)

Nitrogen fixation by *Anabaena azolla* is apparently not adversely affected by the level of combined nitrogen in the water because the organisms are actually sheltered in the leaflet cavity. The *Azolla* must die off and the nitrogen must be mineralized before it becomes available to the plant. This happens when the temperature rises (to as much as 40°–45° C in Indonesia), and the *Azolla* dies and settles to the bottom of the paddy. Nitrogen is released from the decomposing cells and becomes available to the rice plants. The rice plants then turn green and tillering (the production of multiple shoots from the same plant) increases.

Limitations

Use of the *Azolla-Anabaena* association for food production is limited to agricultural soils that can be flooded, and it is best adapted to rice culture. *Azolla* could possibly be used, however, as a nitrogen source for other aquatic plants such as taro (*Colocasia esculenta*) or water chestnut (*Eleocharis dulcis*). In addition, the association may be beneficial in removing nutrients from sewage treatment lagoons. Water, plenty of sunshine, and a temperature regime that favors rhythmic self-destruction of the fern seem to be the most important requirements for success.

Research Needs

- More information is needed about nitrogen-fixation efficiencies of different *Azolla-Anabaena* strain combinations. It is possible that more efficient nitrogen-fixing combinations can be discovered.
- Good husbandry should be developed for using *Azolla-Anabaena* in rice culture systems, as has been done in Indonesia and Vietnam. Technology for using *Azolla* under different soil and climate conditions is needed, particularly for temperate areas.
- Experiments should be made to determine how best to grow *Azolla* with rice to increase both nitrogen fixation and rice yields.

Asymbiotic Fixation

Blue-Green Algae

The blue-green algae are perhaps our most widespread group of nitrogen fixers because they are present almost everywhere on land, in fresh water, and in the sea. Regardless of environmental conditions (except at low pH), there are almost always present some forms that fix atmospheric nitrogen. Blue-green algae fix nitrogen in Antarctic waters as well as in hot springs. They operate over the range of 0°–60°C.

One reason for their wide range of adaptability is that they include many species, each of which may fix nitrogen under varying conditions. Nitrogen-fixing species occur in the genera *Anabaena*, *Aulosira*, *Cylindrospermum*, *Gloeotricha*, *Tolypothrix*, *Calothrix*, *Nostoc*, *Haplosiphon*, and others.

In rice culture, the blue-green algae can be depended on to provide nitrogen consistently. In long-term soil fertility experiments at the International Rice Research Institute (IRRI) in the Philippines, 23 consecutive rice crops have been harvested from soils unfertilized with nitrogen, without any apparent decline in soil nitrogen. The algae replaced the nitrogen removed by the rice crops.

Studies at the Agricultural Research Center in Giza, Egypt, have shown that two blue-green algae, *Tolypothrix tenuis* and *Aulosira fertilissima*, fix more nitrogen than other forms in that region. In a rotation with rice every third year, it has been found advantageous to grow the effective algae and inoculate the rice fields shortly after planting. Of the rice cultured, 10 percent is now inoculated with a dried algal preparation of the two effective cultures and this percentage is expected to increase rapidly in the future. Inocula are also supplied to farmers in India by the Agricultural Research Institute in New Delhi.

The importance of blue-green algae in rice paddies has long been recognized. But these microorganisms also operate very well in desert regions; they use the moisture of night dews during the early morning hours and fix nitrogen during this period of temporary activity. In the western United States blue-green algae in crusts on the soil surface fix considerable amounts of nitrogen per hectare per year. Nitrogenase activity (the activity of the enzyme that splits molecular N_2) ceases when the crusts become dry, but it is measurably reactivated within 2 hours after crusts are moistened.

The importance of the role of the blue-green algae in fixing nitrogen was not appreciated until the acetylene-reduction technique of measuring nitrogen fixation was developed. Now it is possible to study algal fixation in streams and lakes and under various soil conditions, and the significance of the blue-green algae in our world food production is becoming more evident. Their main assets are 1) the wide range of adaptation to temperature and moisture, and 2) their ability to respond quickly when environmental conditions are suitable and to grow rapidly in paddy fields. The blue-green algae fill ecological niches left by other systems of biological nitrogen fixation.

Limitations

From a management standpoint, knowledge of how to use the blue-green algae effectively is meager. Nitrogen fixed by these microorganisms is not readily available to food crops—the algal cells must decompose and the nitrogen must be mineralized. Although this may not be a problem in continuous rice culture, it could be a major obstacle in other planting systems.

Algal growth in freshwater lakes is usually undesirable because they cause the water to become stagnant. Technology on use of algal tissue as feed for livestock or food for human consumption is needed. At present, its use is chiefly as a green manure.

Research Needs

- A survey should be made to determine the occurrence of good nitrogen-fixing species.
- Strains should be selected for environmental adaptation as well as nitrogen-fixing potential.
- Technology related to husbandry and how to prepare, store, and distribute inoculants is needed.
- Methods of culturing starter cultures and distributing dependable inocula should be developed; biological and chemical methods of control will result in better husbandry and more efficient handling of effective algae.
- Information is needed on how to encourage the growth of desirable organisms and discourage the growth of undesirable ones. The role of predators in reducing algal nitrogen fixation should be investigated and the use of algae adapted to local soils.
- Studies should be conducted on the use of algae for human food and animal feed. The large biomass of algae could possibly provide high-quality edible protein for animal consumption.

Free-Living Nitrogen-Fixing Bacteria

Free-living nitrogen fixers from at least 25 genera and many taxonomic groups of bacteria are known. These organisms occur in diverse habitats; their requirements for oxygen, a specific energy source, electron acceptors, and other factors vary widely.

Fixation of significant amounts of nitrogen is dependent upon a suitable carbon and energy supply. An adequate source of energy is one of the most critical limiting factors to nitrogen fixation by free-living organisms. One of three systems of obtaining energy may be utilized:

- Energy may be obtained through breakdown of plant residues in soil. Only rich, fertile soils harbor organic residues in amounts sufficient to provide significant energy. *Clostridium*, *Klebsiella*, and most *Azotobacter* species rely on this source.

- Certain bacteria are favored by root exudates of some plant genotypes that are very efficient photosynthesizers. Exudates from the roots are selective energy sources for particular bacteria. This relationship is often referred to as an "associative symbiosis"; a recent international symposium on nitrogen fixation recommended that it should be termed "biocoenosis." *Azotobacter paspali*, *Beijerinkia* sp., and *Spirillum lipoferum* fit into this category. (Recent work has suggested that *Spirillum lipoferum* should be reclassified as two species of *Azospirillum*: *A. lipoferum* and *A. brasilense* [Krieg and Tarrand, 1977].)
- Other bacteria carry out photosynthesis themselves. But requirements for growth are so restrictive that these microorganisms are not considered highly important agronomically. *Rhodospirillum rubrum* and other photosynthetic bacteria are in this group.

The presence of nitrogen-fixing bacteria in the root zone does not assure that they are actively fixing nitrogen; it does indicate the capability for nitrogen fixation if there is sufficient energy and other conditions are present for growth. If ammonia or nitrates are present in the soil, the organisms will use these to produce new cells and will conserve energy in preference to fixing nitrogen. Energy used for nitrogen fixation cannot be used in reproduction.

The real nitrogen contribution of free-living nitrogen fixers to the soil is uncertain. With a mixed population under natural conditions, it is difficult to assess the contributions of single species.

Clostridium, *Klebsiella*, and several *Enterobacter* species are credited with substantial nitrogen fixation when energy-rich soils are flooded, but the identities of the major aerobic genera are uncertain. *Azotobacter* species, on the other hand, prefer a moist, aerated environment, but they, too, are dependent upon an adequate source of carbon. In both cases, large amounts of energy-rich materials are required if significant amounts of nitrogen are to be fixed. Nitrogen fixation efficiency is low. It takes the equivalent of about 50 kg of sucrose for *Azotobacter* species to fix 1 kg of nitrogen at the oxygen concentrations found in air.

The inoculation of soils and seeds of nonleguminous plants with preparations of *Azotobacter chroococcum* has been practiced in Russia and India for many years. Some types of *Azotobacter* have been credited with increasing crop yields as a result of the nitrogen they fixed, but the low concentration of cells in the soil could not have fixed appreciable amounts of nitrogen.

With the new highly sensitive techniques for measuring nitrogen fixation, some doubt has been cast on the real contribution of free-living bacteria to soil nitrogen. Bacterial inoculation experiments rarely show yield increases as great as the 10 percent level required to attribute statistical significance to the results. The greatest increases are on very fertile soil. There is little probability that any sound inoculation practice for free-living bacteria will be developed soon, and many of the reports of alleged growth stimulation are now considered dubious.

In highly fertile soils, *Azotobacter* species are sometimes believed to produce growth factors and vitamins that are beneficial to vegetables. Benefits from *Azotobacter* inoculation often are attributed to these growth factors rather than to nitrogen fixation. The effect is manifested only in the highly fertile garden soils used in vegetable production. The results of these studies are still equivocal.

The discovery that the association of Bahia grass (*Paspalum notatum*) with *Azotobacter paspali* in tropical soils resulted in nitrogen fixation stimulated new interest in this field. Interest was heightened by the subsequent finding by Dobereiner in Brazil that an associative symbiotic relationship between *Digitaria decumbens*, cultivar "transvala," and the microorganism *Azospirillum lipoferum* also brought about nitrogen fixation. It was reasoned that tropical grasses with their more efficient 4-carbon photosynthetic cycle could indeed provide the abundance of energy needed for fixation that was lacking in other systems.

The enthusiasm has been dampened somewhat by the great variability that has characterized field studies, and more study is needed to identify the limiting factors and devise agronomic practices to bypass them. Large variations in growth cycles are observed with cereals, and they appear to fix nitrogen only during the reproductive phase. Interactions of nitrogen fixation, nitrate assimilation, and denitrification raise the question whether the nitrogen is being lost as rapidly as it is being fixed.

On the other hand, nitrate reductase-negative mutants of *Azospirillum* species are now available that fix nitrogen in the presence of high levels of nitrate. Much more study is needed to identify the factors important for vigorous fixation and to reduce the high degree of variability.

Limitations

Too little is known of the physiology of this unique associative symbiosis for it to be used effectively. In some cases, for instance, the organism may enter the root cortex, but fail to proliferate enough to effect significant nitrogen fixation. We need to know the reason for this.

To date, firm data have not been published to establish that *Azospirillum* and free-living nitrogen fixers contribute substantial amounts of nitrogen and increase crop yields under field conditions.

The conditions required for good inoculation trials are difficult to attain. These conditions are: 1) low numbers of microorganisms already present in the root zone with dinitrogen fixing capability; 2) inoculum able to compete with established root-zone bacteria; 3) soil conditions that will support proliferation of the organism to produce a large biomass of cells; 4) low levels of available combined nitrogen in the soil; and 5) adequate substrate or plant exudate to supply the energy required for fixation.

Research Needs

- More knowledge is needed on the physiology of free-living, nitrogen-fixing bacteria. Only a few strains of the microorganisms, and even fewer genotypes of the host plants, have been studied. A report on nitrogen fixation in wheat, *Triticum aestivum*, is of particular interest. Roots from lines containing the 5-D chromosome were covered with a gelatinous material (probably a polysaccharide) which favored proliferation and nitrogen fixation by gram-positive bacteria within the gelatinous layer on the roots in contrast to wheat lines without this chromosome.
- Mass screening for nitrogen-fixing activity of numerous grass genotypes and strains of microorganisms should prove rewarding.
- Greater emphasis should be placed on field studies. Acetylene-reduction tests should be used when needed, but in field studies, increased yields and higher quality are of greater significance than the amount of nitrogen fixed.
- Attempts should be made to modify the rhizosphere bacterial community to allow the inoculum strain to become established.

References and Suggested Reading

Rhizobium-Leguminous Plant Associations

Brill, W. J. 1977. Biological nitrogen fixation. *Scientific American* 236:68-74.

Buchanan, R. E., and Gibbons, N. E., eds. 1974. *Bergey's manual of determinative bacteriology*. 8th edition. Baltimore: The Williams and Wilkins Co.

Burns, R. C., and Hardy, R. W. F. 1975. *Nitrogen fixation in bacteria and higher plants.* Berlin: Springer-Verlag.

Burris, R. H. 1975. The acetylene reduction technique. In *Nitrogen fixation by free-living microorganisms:* International Biological Programme 6, pp. 249-257. Cambridge, England: Cambridge University Press.

Burton, J. C. 1967. Rhizobium culture and use. In *Microbial technology*, H. J. Peppler, ed., pp. 1-33. Huntington, New York: Robert E. Krieger Publishing Co.

Evans, H. J. 1969. How legumes fix nitrogen. In *Crops grown–a century later*, Agricultural Experiment Station Bulletin No. 708, pp. 110-127. New Haven: Connecticut Agricultural Experiment Station.

____ . 1975. *Enhancing biological nitrogen fixation: proceedings of a workshop held on June 6, 1974.* Sponsored by Energy Related Research and the Division of Biological and Medical Sciences of the National Science Foundation. Washington, D.C.: U.S. National Science Foundation.

Fred, E. B.; Baldwin, I. L.; and McCoy, E. 1932. *Root-nodule bacteria and leguminous plants*. Madison: University of Wisconsin Press.

Hardy, R. W. F., and Havelka, U. D. 1970. Nitrogen fixation research, a key to world food. *Science* 188:633-643.

Silver, W. S., and Hardy, R. W. F. 1976. *Biological nitrogen fixation in forage and livestock systems.* American Society of Agronomy Special Publication No. 28, pp. 1-34. Madison, Wisconsin: American Society of Agronomy.

Skinner, K. J. 1976. Nitrogen fixation—key to a brighter future for agriculture. *Chemical and Engineering News* 54:22-35.

Frankia-Nonleguminous Plant Associations

Allen, E. K., and Allen, O. N. 1964. Non-leguminous plant symbiosis. In *Microbiology and soil fertility*, 25th Annual Biology Colloquium, C. M. Gilmour and O. N. Allen, eds., pp. 77-106. Corvallis: Oregon State University Press.

Becking, J. H. 1977. Dinitrogen-fixing associations in higher plants other than legumes. In *A treatise on dinitrogen fixation*, R. W. F. Hardy and W. Silver, eds., Section III: *Biology*, pp. 185-276. New York: John Wiley and Sons.

Bond, G. 1974. Root-nodule symbiosis with actinomycete-like organisms. In *The biology of nitrogen fixation*, A. Quispel, ed., pp. 342-378. Amsterdam: North-Holland Publishing Co.

Callaham, D.; Tredici, P. D.; and Torrey, J. G. 1978. Isolation and cultivation *in vitro* of the actinomycete causing root nodulation in *Comptonia. Science* 199:899-902.

Silvester, W. B. 1977. Dinitrogen fixation by plant associations excluding legumes. In *A treatise on dinitrogen fixation*, R. W. F. Hardy and A. H. Gibson, eds., Section IV: *Agronomy and ecology*, pp. 141-190. New York: John Wiley and Sons.

Torrey, J. G. 1978. Nitrogen fixation by actinomycete-nodulated angiosperms. *BioScience* 28:586-592.

Azolla-Anabaena Associations

Becking, J. H. 1975. Contribution of plant-algae associations. In *Proceedings of the International Symposium on Nitrogen Fixation*, pp. 556-580. Pullman: Washington State University Press.

Mague, T. H. 1977. Ecological aspects of dinitrogen fixation by blue-green algae. In *Treatise on dinitrogen fixation*, R. W. F. Hardy and A. H. Gibson, eds., Section IV: *Agronomy and ecology*, pp. 85-140. New York: John Wiley and Sons.

Moore, A. W. 1969. *Azolla*: biology and agronomic significance. *Biological Review* 35:35-37.

Peters, G. A. 1975. Studies on the *Azolla: Anabaena* symbiosis. In *Proceedings of the International Symposium on Nitrogen Fixation*, W. E. Newton and C. J. Nyman, eds., pp. 592-610. Pullman: Washington State University Press.

____ . 1978. Blue-green algae and algal associations. *BioScience* 28:580-585.

Blue-Green Algae

Burris, R. H. 1975. The acetylene-reduction technique. In *Nitrogen fixation by free-living microorganisms.* International Biological Programme 6, pp. 249-257. Cambridge, England: Cambridge University Press.

Dart, P. J., and Day, J. M. 1977. Non-symbiotic nitrogen fixation in soil. In *Soil microbiology*, N. Walker, ed., pp. 225-252. New York: John Wiley and Sons.

Fogg, G. E. 1971. Nitrogen fixation in lakes. In *Plant and soil special volume: biological nitrogen fixation in natural and agricultural habitats.* Proceedings of the Technical Meetings on Biological Nitrogen Fixation of the International Biological Program (Section PP-N), Prague and Wageningen, 1970, T. A. Lie and E. G. Mulder, eds., pp. 393-401. The Hague: Martinus Nijhoff.

Hendrikkson, E. 1971. Algae nitrogen fixation in temperate regions. In *Plant and soil special volume: biological nitrogen fixation in natural and agricultural habitats.* Proceedings of the Technical Meetings on Biological Nitrogen Fixation of the International Biological Program (Section PP-N), Prague and Wageningen, 1970, T. A. Lie and E. G. Mulder, eds., pp. 415-419. The Hague: Martinus Nijhoff.

Rinaudo, G.; Balandreau, J.; and Dommergues, Y. 1971. Algae and bacterial non-symbiotic nitrogen fixation in paddy soils. In *Plant and soil special volume: biological nitrogen fixation in natural and agricultural habitats.* Proceedings of the Technical Meetings on Biological Nitrogen Fixation of the International Biological Program (Section PP-N), Prague and Wageningen, 1970, T. A. Lie and E. G. Mulder, eds., pp. 471-479. The Hague: Martinus Nijhoff.

Stewart, W. D. P. 1966. Nitrogen fixation by free-living organisms. In *Nitrogen fixation in plants*, pp. 68-83. London: Athlone Press. Distributed in the United States by Humanities Press, Atlantic Highlands, New Jersey.

_____, ed. 1976. Blue-green algae. *Nitrogen-fixation by free-living micro-organisms.* International Biological Programme Series 6, pp. 129-229. Cambridge, England: Cambridge University Press.

Free-Living Nitrogen-Fixing Bacteria

Barber, L. E.; Tjepkema, J. D.; Fussell, S. A.; and Evans, H. J. 1976. Acetylene reduction (nitrogen fixation) associated with corn inoculated with *Spirillum. Applied and Environmental Microbiology* 32:108-113.

Burris, R. H.; Albrecht, S. L.; and Okon, Y. 1978. Physiology and biochemistry of *Spirillum lipoferum.* In *Limitations and potentials for biological nitrogen fixation in the tropics*, Vol. 10, Basic Life Sciences, Proceedings of a Conference on Limitations and Potentials of Biological Nitrogen Fixation in the Tropics, Brasília, Brazil, Johanna Dobereiner, Robert H. Burris, Alexander Hollaender, Avilio A. Franco, Carlos A. Neyra, and David Barry Scott, eds., pp. 303-315. New York: Plenum Press.

Dart, P. J., and Day, J. M. 1975. Nitrogen fixation in the field other than by nodules. In *Soil microbiology: a critical view*, Norman Walker, ed., pp. 225-252. London: Butterworth's Scientific Publications.

Knowles, R. 1977. The significance of asymbiotic dinitrogen fixation by bacteria. In *A treatise on dinitrogen fixation,* R. W. F. Hardy and A. H. Gibson, eds., Section IV: *Agronomy and ecology*, pp. 33-84. New York: John Wiley and Sons.

Krieg, N. R., and Tarrand, J. J. 1977. Taxonomy of the root-associated nitrogen fixing bacterium *Spirillum lipoferum.* In *Limitations and potentials for biological nitrogen fixation in the tropics*, Vol. 10, Basic Life Sciences, Proceedings of a Conference on Limitations and Potentials of Biological Nitrogen Fixation in the Tropics. Brasília, Brazil. Johanna Dobereiner, Robert H. Burris, Alexander Hollaender, Avilio A. Franco, Carlos A. Neyra and David Barry Scott, eds., pp. 317-333. New York: Plenum Press.

Research Contacts

Rhizobium-Leguminous Plant Associations

R. H. Burris, Department of Biochemistry, University of Wisconsin, Madison, Wisconsin 53706, U.S.A.

R. A. Date, CSIRO, The Cunningham Laboratory, Mill Road, St. Lucia, Queensland, Australia, 4067.

Deane F. Weber, Cell Culture and Nitrogen Fixation Laboratory, Beltsville Agricultural Research Center, U.S. Department of Agriculture, Beltsville, Maryland 20705, U.S.A.

Frankia-Nonleguminous Plant Associations

J. H. Becking, Institute for Atomic Sciences in Agriculture, 6 Keyenbergseweg, Postbus 48, Wageningen, The Netherlands.

W. B. Silvester, Department of Biological Sciences, University of Waikato, Private Bag, Hamilton, New Zealand.

Azolla-Anabaena Associations

Alan W. Moore, CSIRO, The Cunningham Laboratory, Mill Road, St. Lucia, Queensland, Australia, 4067.

D. W. Rains, Plant Growth Laboratory, University of California, Davis, California 95616, U.S.A.

Blue-Green Algae

T. M. Mague, Bigelow Laboratory for Ocean Sciences, McKown Point, West Boothbay Harbor, Maine 04575, U.S.A.

W. D. P. Stewart, Department of Biological Sciences, University of Dundee, Dundee, DD1 4HN, Scotland.

I. Watanabe, International Rice Research Institute, Los Baños, The Philippines.

Free-Living Nitrogen-Fixing Bacteria

Lynn Barber, Department of Microbiology, Oregon State University, Corvallis, Oregon 97331, U.S.A.

R. H. Burris, Department of Biochemistry, University of Wisconsin, Madison, Wisconsin 53706, U.S.A.

Johanna Dobereiner, EMBRAPA, 23460 Seropedica, Rio de Janeiro, Brazil.

David H. Hubbell, Soil Science Department, University of Florida, Gainesville, Florida 32611, U.S.A.

Chapter 5

Microbial Insect Control Agents

There are more than 1,500 naturally occurring microorganisms or their products that hold promise for the control of major insect pests. Microorganisms that affect insects are termed entomopathogens. They may be used to induce diseases in target insects or to suppress populations of insects directly or in combination with chemical insecticides.

Energy-efficient pest-control approaches must be developed to reduce to a minimum the use of factory-produced, toxic, broad-spectrum chemical insecticides used in industrial countries in ever-increasing amounts. The microbial approach can be applied to agricultural practices of both developing and developed countries, and many of its techniques are ready for implementation. By adopting a systems approach to "integrated pest management," using entomopathogens and other nonchemical factors for specific pests, developing countries have an important opportunity to bypass the traditional chemical approach to insect control.

All types of microorganisms are represented among the potential microbial control agents. As an example, nearly 100 species of bacteria and over 700 viruses have been isolated from arthropods and more are being discovered each year. All classes of fungi are represented among the more than 750 known entomopathogenic fungi. Protozoa are also likely candidates as microbial control agents because many insects not attacked by other entomopathogens are susceptible to at least one of the 300 known species of entomophilic protozoa.

Development of Bioinsecticides

In planning an approach to the use of microbial control agents, the most significant factors to be considered include production technology, safety and specificity, and efficacy.

Production Technology

Entomopathogens are produced by fermentation methods (Figure 5.1), in living insects (Figure 5.2), and in cell tissue cultures. Fermentation technology is used for some bacteria and fungi, whereas living insects can be used for the production of obligatory parasitic viruses and protozoa. Both processes have been successfully used to produce commercial entomopathogenic products. For example, submerged fermentation is generally used for commercial production of the entomogenous bacteria *Bacillus thuringiensis* and *Bacillus moritai* and the fungi *Beauveria bassiana* and *Entomophthora virulenta.* Surface fermentation is employed to produce pathogenic fungi, for example, *Nomuraea rileyi* and *Metarrhizium anisopliae*, and a combination of both surface and submerged techniques for *Beauveria bassiana* and *Hirsutella thompsonii*. Living insects are exclusively used as substrates for the production of the respective nucleopolyhedrosis viruses of *Heliothis zea, Porthetria dispar*, and *Hemerocampa pseudosugata.* Cell tissue culture methods currently produce only small quantities of viruses, but they are considered to be the mass production method of the future.

Safety and Specificity

Entomopathogens are infectious, replicating living organisms that are a natural part of our environment. Evidence that microbial insecticides pose little human or environmental hazard has been demonstrated by laboratory animal testing data developed to support federal pesticide registration. Nevertheless, safety cannot be absolutely guaranteed for all entomopathogens in every living system, and it is important that potential hazards for new entomopathogens be known prior to use. Informal guidance for evaluating the specificity and risks in the use of microbial agents has been provided by regulatory agencies. Formal guidelines are now being developed by the U.S. Environmental Protection Agency (EPA). Several baculoviruses have been tested in living organisms with no evidence of toxic or pathogenic effects on vertebrates or nontarget invertebrates.

Baculoviruses do not appear to replicate in vertebrate embryos or in cell lines derived from birds, fishes, amphibians, or mammals. No deleterious effects at normal field-use rates were reported in tests with such bacteria as *Bacillus thuringiensis*, *Bacillus popilliae*, and *Bacillus moritai.* While allergens are encountered among the fungi (*Beauveria*, *Entomophthora*, *Hirsutella*, *Metarrhizium*, and *Nomuraea* species), results indicate that these fungi are not toxic or infectious to vertebrates. Three entomopathogenic protozoa, one from grasshoppers (*Nosema locustae*), one from mosquitoes (*Nosema algerae*), and one from beetles (*Mattesia trogodermae*), have been evaluated against nontarget organisms. Initial *in vivo* tests indicate no apparent risk to vertebrates.

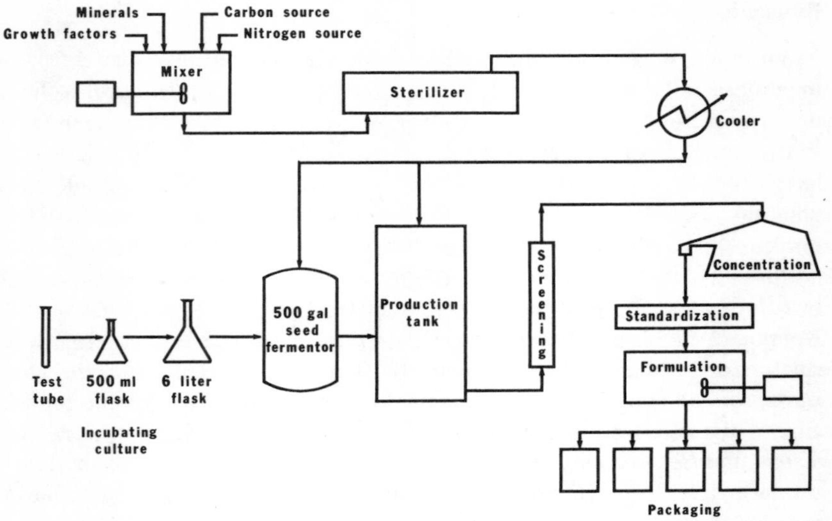

FIGURE 5.1 Submerged fermentation pathway used to produce *Bacillus thuringiensis.* (Photograph courtesy of C. M. Ignoffo)

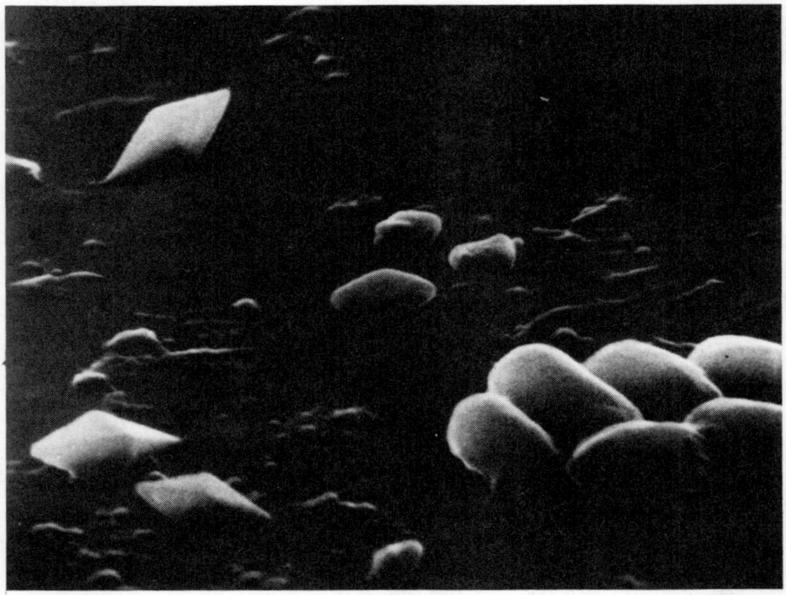

FIGURE 5.2 Scanning electron micrograph of entomocidal parasporal crystals and spores of *Bacillus thuringiensis.* (Photograph courtesy of C. M. Ignoffo)

Efficacy

Entomopathogens have been used to control mites, beetles, and caterpillar pests of agricultural crops, forests, and stored products with varying degrees of success.

Microbial insecticides, like chemical insecticides, are usually sprayed or dusted on crops. Entomopathogens may also be successfully introduced and established in an ecosystem by other application methods to provide long-term control of pest populations. For example, insects themselves can be used to disseminate entomopathogens. Virus or fungus epizootics might be induced in an insect population before crop-damaging proliferation takes place. It may also be possible to manipulate the environment to create conditions in which naturally occurring pathogens exert their greatest effect. Some of these approaches may provide levels of control equal to or better than those currently obtained with chemical insecticides, but further research is needed to exploit their potential.

The potential for substituting microbial control agents for chemical pesticides can be deduced from the following examples in the United States shown in Table 5.1.

Development of the use of entomopathogens or their by-products for microbial control agents is underexploited. Safe, effective entomopathogens formulated as microbial control agents are being developed by governmental agencies and industry, and the commercial products are being effectively used by growers. The newer agents have not been brought to their fullest potential.

TABLE 5.1 Potential Substitution of Chemical Pesticides by Microbial Control Agents in the United States

Disease	Control Agent	Potential Replacement of Chemical Pesticide (t per year)
Cotton bollworm and budworm *(Heliothis zea)*	*Baculovirus heliothis*	>7,700
Citrus rust mite (Florida) *(Phyllocoptruta oleivora)*	*Hirsutella thompsonii*	1,800–7,200
Cabbage looper (Florida) *(Trichoplusia ni)*	*Bacillus thuringiensis*	450–1,400
Western-range grasshoppers	*Nosema locustae*	450– 900
Green peach aphid on Maine potatoes *(Myzus persicae)*	*Entomophthora ignobilis*	90– 310

Bacteria

Many bacteria are associated with insects, most of which belong to the families Pseudomonadaceae, Enterobacteriaceae, Lactobacillaceae, Micrococcaceae, and Bacillaceae (Table 5.2). Members of these families may be obli-

TABLE 5.2 Examples of Bacteria Pathogenic for Insects*

Bacteria	Insects
Pseudomonadaceae	
**Pseudomonas aeruginosa*	Grasshoppers (Orthoptera)
Pseudomonas septica	Scarab beetles (Scarabaeidae), striped ambrosia beetle (*Tripodendron lineatum*)
Vibrio leonardia	Greater wax moth (*Galleria melonella*), European corn borer (*Ostrinia nubilalis*)
Enterobacteriaceae	
Serratia marcescens and	Varieties of butterfly, moth and skipper (Lepidoptera)
**Escherichia coli*	Grasshoppers (Orthoptera), varieties of butterfly, moth and skipper (Lepidoptera)
Enterobacter aerogenes	Grasshoppers (Orthoptera)
Proteus vulgaris, P. mirabilis, and *P. retigeri*	Honeybees (Apidae), greater wax moth (*Galleria melonella*)
**Salmonella schottmuelleri* var. *alvei*	Greater wax moth (*Galleria melonella*)
**Salmonella enteritidis,* **Shigella dysenteriae*	
Lactobacillaceae	
Diplococcus and *Streptococcus* spp.	Cockchafer (*Melolontha melolontha*), silkworm (*Bombyx mori*), gypsy moth (*Lymantria dispar*), processionary moths (*Thaumetopoeia* spp.)
**Streptococcus faecalis*	Greater wax moth (*Galleria melonella*)
Micrococcaceae	
**Micrococcus* spp.	Green June beetle (*Cotinis nitida*), sawflies (Tenthredinidae), houseflies (Muscidae), various Lepidoptera including nun moth (*Lymantria monacha*), European corn borer (*Ostrinia nubilalis*), and cutworms (Noctuidae)
Bacillaceae	
Bacillus thuringiensis and *B. cereus*	Varieties of butterfly and moth (Lepidoptera)
Bacillus popilliae and *B. lentimorbus*	Scarab beetles (Scarabaeidae)
Bacillus sphaericus	Mosquitoes (Culicidae)
Bacillus larvae	Honeybees (Apidae)
Bacillus moritai	Flies (Diptera)
Clostridium novyi and *C. perfringens*	Greater wax moth (*Galleria melonella*)

*Those asterisked may also be pathogenic for man.

gate or opportunistic entomopathogens, depending on their host association in nature. Obligate entomopathogens are generally fastidious and are restricted to growth in a living host insect. The occasional or opportunistic pathogens are free living in nature, although they may commonly be found associated with one or more hosts. About 100 bacteria have been reported as entomopathogens, but only four (*Bacillus thuringiensis*, *B. popilliae*, *B. lentimorbus*, and *B. sphaericus*) have been closely examined as insect-control agents. These species are sporeformers: the first two produce, in addition to the spore, discrete crystalline inclusions within the sporulating cell; the last two do not.

Production

Fermentation processes are employed in the production of *B. thuringiensis* and *B. sphaericus* spores, whereas *B. popilliae* or *B. lentimorbus* spores are produced exclusively in living insects. Two types of fermentation methods are used for the production of *B. thuringiensis* and *B. sphaericus*: 1) surface or semisolid, and 2) submerged. In surface fermentation, the *B. thuringiensis* spores are inoculated on a semisolid medium composed of wheat bran, expanded perlite, soybean meal, glucose, and inorganic salts. The bran is harvested after 36–48 hours and formulated to a product active against insects. Submerged fermentation of *B. thuringiensis* (Figure 5.1) is carried out in a liquid medium using standard fermentation protein sources (fish meal, soy flour) and carbohydrate sources (molasses, sucrose). After sporulation of the bacteria, the spent liquor is passed through a fine screen, the active ingredients are centrifuged from the medium, mixed with stabilizing agents, and then packaged as either a powder or liquid. Similar materials may be used for *B. sphaericus.*

For the *in vivo* production of *B. popilliae* and *B. lentimorbus*, third instar Japanese beetle (*Popillia japonica*) larvae are infected with spores and incubated in soil seeded with rye to feed the larvae. The infected larvae are harvested 16–21 days after infection. The infected larvae are pulverized, stabilizing agents are added, and the slurry is dried and packaged as a dry powder. Living insects must be used to propagate these bacteria because artificial culture methods that will support large-scale sporulation of the *E. popilliae* and *B. lentimorbus* are not as yet available.

Safety and Specificity

B. thuringiensis, *B. popilliae*, *B. lentimorbus*, and *B. sphaericus* have been subjected to many safety tests, with no harmful effects for animals or human beings.

What impact these bacteria have on the environment when used as insecticides is difficult to predict, because little data are available on persistence and concentration of bacterial agents in the food chain. True parasites are obviously dependent for existence on the population of their hosts.

Efficacy

B. thuringiensis (Figure 5.2) is one of the best-known and most widely used microbial control agents. It is pathogenic for lepidopteran larvae (Figure 5.3) affecting more than 150 larval species that include some of the most important economic pests listed in Table 5.3. Preparations of *B. thuringiensis* can be mixed with a number of commercial insecticides, fungicides, and various adhesives and wetting agents. Commercial products are applied to field crops and stored commodities, employing the same methods used for chemical pesticides.

B. popilliae and *B. lentimorbus* are pathogens of various beetles. When ingested by beetle larvae, the bacteria invade the blood system, where they proliferate and sporulate, causing death of the larvae. The mass of spores that accumulates prior to death of the insect is ultimately released and survives for extended periods in the soil. These spores may be eaten by newly hatched beetle larvae and, upon germination and growth in the insect gut, begin the infectious process again. The name given to this infection is "milky disease" because the infected larvae take on a whitish or milky appearance. One appli-

FIGURE 5.3 Diseased larvae of the cabbage looper with symptoms of *B. thuringiensis* poisoning. (Photograph courtesy of C.M. Ignoffo)

TABLE 5.3 Some Insect Pests Susceptible to Control with Preparations of *Bacillus thuringiensis*

Insect Pest	Plants Affected
Cabbage looper (*Trichoplusia ni*)	Broccoli, cabbage, cauliflower, celery, lettuce, potato, melon
Imported cabbageworm (*Pieris rapae*)	Broccoli, cabbage, cauliflower
Tobacco hornworm (*Manduca sexta*)	Tobacco
Tobacco budworm (*Heliothis virescens*)	Tobacco
Tomato hornworm (*Manduca quinquemaculata*)	Tomato
Alfalfa caterpillar (*Colias eurytheme*)	Alfalfa
Gypsy moth (*Lymantria dispar*)	Forest trees
European corn borer (*Ostrinia nubilalis*)	Corn
Grape leaffolder (*Desmia funeralis*)	Grape
Codling moth (*Laspeyresia pomonella*)	Apples, pears
Green cloverworm (*Plathypena seabra*)	Soybeans
Orangedog (*Pailio cresphontes*)	Citrus
Range caterpillar (*Hemileuca oliviae*)	Range grass
Sugarcane borer (*Diatraea saccharalis*)	Sugarcane
Cotton bollworm (*Heliothis zea*)	Cotton
Spruce budworm (*Choristoneura fumiferana*)	Forest trees
Indian meal moth (*Plodia interpunctella*)	Stored grains

cation of these bacteria lasts many seasons, although the physical and chemical properties of the soil as well as climatic conditions, agricultural practices, and larval population density influence their effectiveness in nature over prolonged periods.

Several strains of *B. sphaericus* have been isolated that are highly toxic and specific to larvae of disease-carrying mosquitoes. The infectivity of *B. sphaericus* strains varies with the mosquito species; in general, larvae of the genus *Aedes* are the least susceptible, while those of the genus *Culex* are the most susceptible, at least at the current stages of strain development. Experimentally, this bacterium has been produced commercially at prices competitive with those of chemical insecticides. All isolates that have insecticidal activity have not been fully characterized.

Limitations

The basic premise on which the insecticide industry operates and the conditions under which the EPA will issue registrations is that the pesticide must be useful against the target species without adversely affecting man, nontarget animals, or plants. Other parameters to be considered for extensive use of bacteria to control insects are: 1) convenience of application (applicable as a dust, liquid spray, or bait); 2) possibility of integration with conventional chemical insecticides; 3) storage characteristics; 4) economics; 5) ease of production; and 6) safety. The success of a bacterial insecticide, as for any pesticide, also depends on a variety of interacting factors such as environmental and climatic conditions, commodity to be protected (field crop or stored product), mode and timing of application, behavior and habits of the target host(s), and defense mechanism(s) of the insect.

As noted above, field testing of the efficacy of *B. thuringiensis*, *B. popilliae*, *B. lentimorbus*, and *B. sphaericus* has been done and preparations of all these organisms are commercially produced, with the exception of *B. sphaericus*. Because the spore stage is packaged for dissemination, most of the above criteria have been met for *B. thuringiensis*, *B. popilliae*, and *B. lentimorbus*. Although *B. thuringiensis* and *B. sphaericus* meet most of the above criteria, *B. popilliae* and *B. lentimorbus* have some serious drawbacks; they have limited commercial use because of the difficulty of producing spores in quantity.

A much clearer understanding of the metabolism of both the insect host and the bacterial pathogen is needed. For example, to facilitate successful industrial fermentation of spores of *B. popilliae* and *B. lentimorbus,* insight into the mechanisms that control bacterial spore formation during the infectious process is necessary.

Research Needs

The following are some of the most pressing needs for better understanding and use of bacterial insect pathogens:

- To identify new bacterial pathogens;
- To develop *in vitro* production of *B. popilliae* and *B. lentimorbus* and determine nutrients required for sporulation;
- To identify possible biohazards;
- To increase understanding of the mechanisms of insect infection; and
- To encourage industrial research and development through appropriate incentives.

Viruses

Virus diseases have been described for most major arthropod pests. About 700 viruses have been isolated from insects and mites. Most viruses (83 percent of those described in Table 5.4) have been isolated from caterpillars, since many moth and butterfly larvae are serious economic pests. Viruses of flies and sawflies account for about 14 percent of those listed in Table 5.4; the other 3 percent are equally divided among viruses of beetles, grasshoppers, and mites.

Insect viruses fall into five major groups:

- nucleopolyhedrosis viruses (NPV)
- cytoplasmic polyhedrosis viruses (CPV)
- granulosis viruses (GV)
- entomopox viruses (EPV)
- non-inclusion viruses (NIV).

Figure 5.4 shows electron micrographs of these viruses, and Table 5.4 gives examples of some host species that are being considered for control.

The NPV and GV, because of their specificity, safety, virulence, and stability, are probably the most promising candidates for development as viral insecticides. The CPV, EPV, and NIV are not currently considered likely candidates because less is known about their host range, production feasibility, stability, and efficacy.

Production

Insect viruses are strict parasites and must be mass-produced in living hosts or cell cultures. This means rearing the target pest insect and producing the virus by artificial infection, harvesting the virus, and formulating an effective insecticidal preparation.

During early phases of the development of viral insecticides, insects were collected in the field and fed contaminated foliage. Dying insects were then processed into virus preparations for subsequent use. Recently developed techniques and semisynthetic diets permit year-round virus production. For example, bollworms are mass-reared and the bollworm NPV grown in the larvae (Figure 5.5). Tissue culture may be a virus-production technology of the future. Insect tissue cell-culture lines have been established in which insect viruses grow and multiply.

In spite of the highly specialized technology needed to mass-produce viruses, every major group of insect virus has been produced and used in the field as an insecticide. In addition, more than a dozen commercial or experi-

TABLE 5.4 Examples of Viruses and Their Hosts

Nucleopolyhydron Viruses (NPV)	
Alfalfa looper (*Autographa californica*)	
Almond moth (*Ephestia cautella*)	Ermine moth (*Yponomeuta padella*)
Army worm (*Pseudaletia unipuncta*)	Forest tent caterpillar (*Malacosoma disstria*)
Asiatic rice borer (*Chilo supressalis*)	Gypsy moth (*Lymantria dispar*)
Beet armyworm (*Spodoptera exigua*)	Imported cabbageworm (*Pieris rapae*)
Bollworm (*Heliothis zea*)	Mamestra cabbageworm (*Mamestra brassicae*)
Cotton cutworm (Noctuidae)	Pine sawflies (*Neodyprion* spp.)
Cotton leafworm (*Alabama argillacea*)	Spruce budworm (*Choristoneura fumiferana*)
Cotton leaf-perforator (*Buccylatrix thurberiella*)	Tobacco budworm (*Heliothis virescens*)
Cabbage looper (*Tricoplusia ni*)	Variegated cutworm (*Peridroma saucia*)
Corn earworm (*Heliothis zea*)	Wattle bagworm (*Kotochalia jundoi*)
Douglas fir tussock moth (*Orygia pseudotsugata*)	Whitemarked tussock moth (*Orygia leucostigma*)
Granulosis Viruses (GIV)	
Asiatic rice borer (*Chilo supressalis*)	Redbanded leafroller (*Argyrotaenia velutinana*)
Cabbageworm (*Pieris* spp.)	Saltmarsh caterpillar (*Estigmene acrea*)
Codling moth (*Laspeyresia pomonella*)	Spruce budworm (*Choristoneura fumiferana*)
Potato tuberworm (*Phthorimaea operculella*)	
Cytoplasmic Polyhedron Viruses (CPV)	
Cabbageworm (*Pieris* spp.)	Pine caterpillar (*Thaumetopoeia* spp.)
Mamestra cabbageworm (*Mamestra brassicae*)	Pine processionary caterpillar (*Thaumetopoeia* spp.)
Pink bollworm (*Pectinophora gossypiella*)	Spruce budworm (*Choristoneura fumiferana*)
Non-inclusion Viruses (NIV)	
Asiatic rice borer (*Chilo supressalis*)	Coconut rhinoceros beetle (*Xyloryctes jamaicensis*)
Citrus red mite (*Panonychus citri*)	

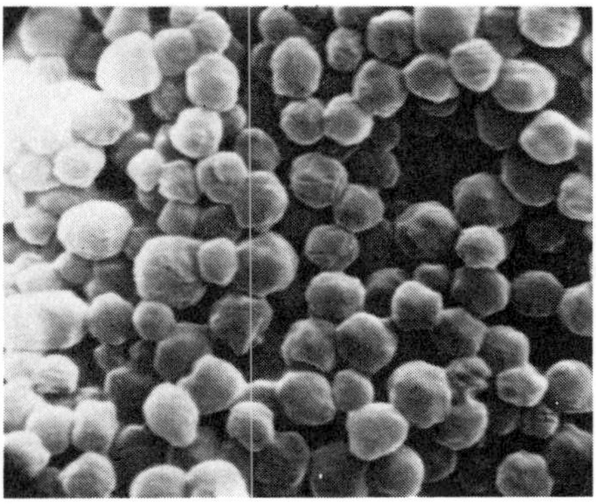

(a) Nucleopolyhedrosis virus. (Photograph courtesy of C. M. Ignoffo)

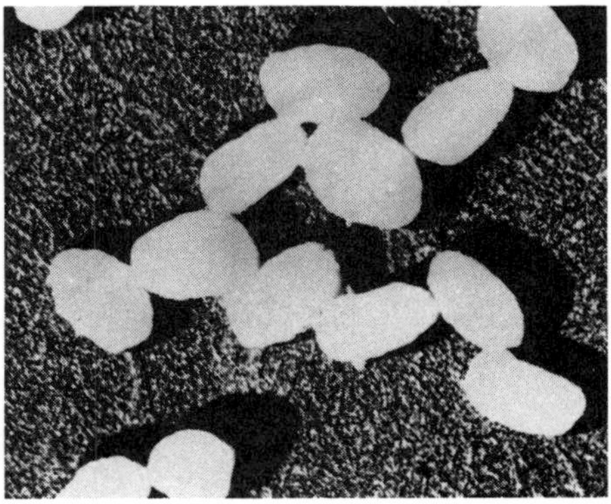

(b) Granulosis virus. (Photograph courtesy of C. M. Ignoffo)

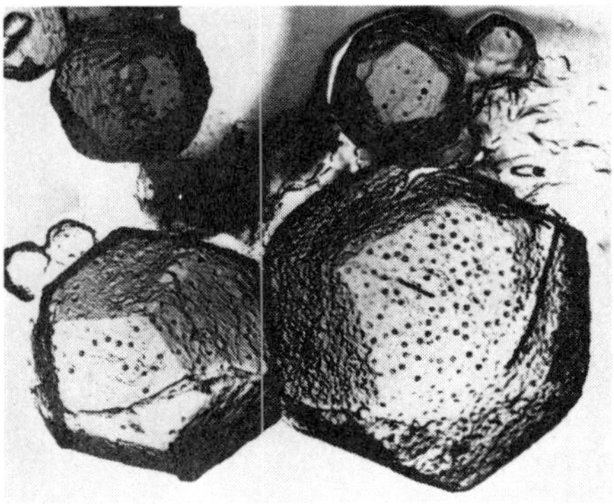

(c) Cytoplasmic polyhedrosis virus. (Photograph courtesy of J. Adams)

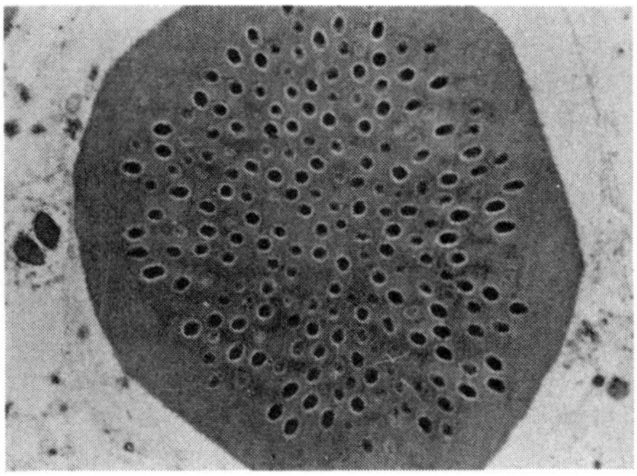

(d) Entomopox virus. (Photograph courtesy of M. Bergoin)

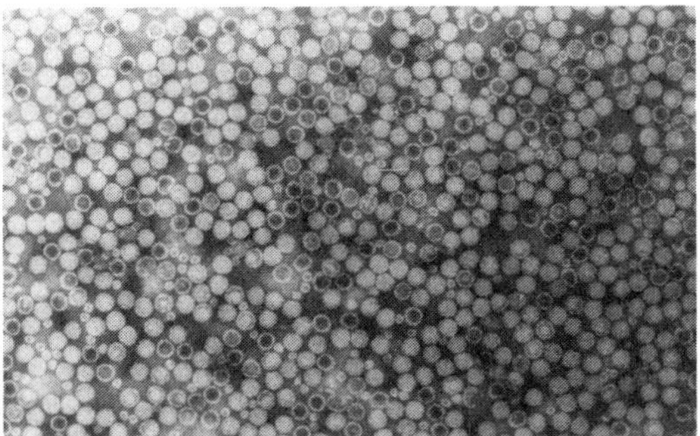

(e) Non-inclusion virus. (Photograph courtesy of C. Vago)

FIGURE 5.4 Electron micrographs of the five types of viruses associated with insects.

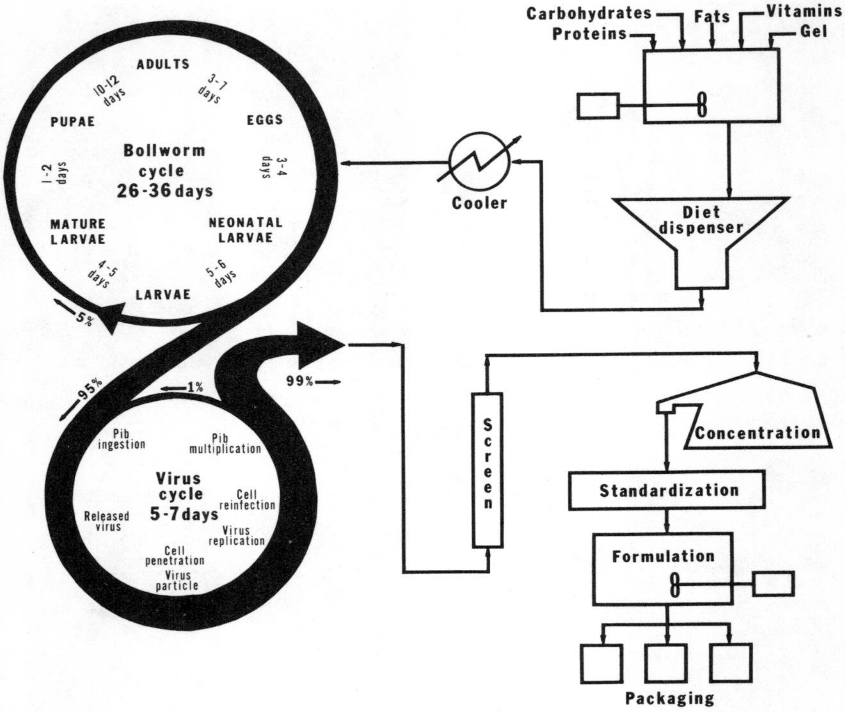

FIGURE 5.5 *In vivo* process used to produce a nucleopolyhedrosis virus. (Photograph courtesy of C. M. Ignoffo)

mental preparations of entomopathogenic viruses have been evaluated sufficiently to warrant designation of a trade name.

Safety and Specificity

Insect viruses should be tested to determine the effects they may have on man, other animals, and plants. There is indirect evidence to suggest that viruses are safe, since man is continually exposed to them on marketed fruits and vegetables, with no apparent ill effects. Further, there is no evidence of toxicity or pathogenicity to those handling the virus in research and production. Such evidence, however, must be corroborated by direct experimentation. Although absolute safety is not a prerequisite for use, testing sufficient to demonstrate that the virus can be used without adverse effect should be completed prior to large-scale field use. Its use can then be monitored for several years to ensure that unforeseen environmental effects do not develop. In the light of present knowledge baculoviruses appear to be safe and specific.

Efficacy

The occurrence of natural epizootics that severely reduce populations of insect pests are striking demonstrations of the effectiveness of insect viruses. These natural epizootics can occur in many insect populations; they are commonly observed in field populations of sawflies, gypsy moths, tent caterpillars, cabbage loopers, and alfalfa caterpillars. Unfortunately, most of this natural control comes after the crop damage occurs. The same control, timed to prevent crop loss, has been achieved by introducing appropriate insect viruses to attack field populations of insects (Table 5.5).

TABLE 5.5 Examples of Viruses Tested for Control of Arthropod Pests.

Crop Attacked and Target Pest	Virus Type
Vegetables	
Cabbage looper (*Trichoplusia ni*)	Nucleopolyhedrosis
Cabbageworm (*Pieris* spp.)	Granulosis
Orchard Crops	
Citrus red mite (*Panonychus citri*)	Non-inclusion
Codling moth (*Laspeyresia pomonella*)	Granulosis
Forage Crops	
Alfalfa caterpillar (*Colias eurytheme*)	Nucleopolyhedrosis
Fiber and Grain Crops	
Boll- and budworms (*Heliothis zea*)	Nucleopolyhedrosis
Cotton leafworm (*Alabama argillacea*)	Nucleopolyhedrosis
Forest and Shade Trees	
European pine processionary moth (*Thaumetopoeia* spp.)	Cytoplasmic polyhedrosis
Gypsy moth (*Lymantria dispar*)	Nucleopolyhedrosis
Spruce budworm (*Choristoneura fumiferana*)	Entomopox

Limitations

Viruses must be grown in living systems, which means that a technology for mass production must be developed for each type of virus. Each virus must then be individually evaluated to determine its specificity and its potential environmental impact before field use is possible.

Although protocols and guidelines are available for evaluating the risks of using insect viruses, only a few viruses (Baculoviruses) have been evaluated in this way.

Some viruses have a short residual field life, which may ultimately reduce efficacy against a target pest. Few can be shown to persist in nature as reservoirs of infection apart from their hosts.

Long-term storage may be a problem, especially at higher temperatures.

Viruses must infect and grow in insects to be effective. Because they are relatively host-specific, they infect only one insect or those several closely related. This is an advantage from the viewpoint of environmental effects.

As with all insecticides, thorough coverage of an infested area is required for effective control.

Research Needs

The more important research and development needs related to the use of viruses as bioinsecticides include the following:

- Compiling and maintaining a repository of potential viruses for crop and forest and of vectors of diseases of man and animals;
- Developing better techniques for the identification and detection of viruses;
- Increasing the understanding of how insect viruses infect and replicate in their hosts;
- Developing more definitive criteria for evaluating the possible biohazards of entomopathogens;
- Improving technology for production, formulation, and use of viral entomopathogens; and
- Encouraging industrial participation in the development of entomopathogens with subsidies, grants, licensing agreements, patent protection, and similar incentives.

Protozoa

Protozoa are primarily free-living, but many types are intimately associated with insects in relationships ranging from compatible to harmful. Most of the approximately 300 described species of entomophilic protozoa are included in the orders Microsporidia (Subphylum Cnidospora) and Neogregarinida (Subphylum Sporozoa).

Microsporidia, particularly common in insects, generally possess the greatest potential as entomopathogens (Table 5.6). Among the microsporidia,

TABLE 5.6 Protozoa as Promising Candidates for Insect Control Agents

Protozoa	Insect Host(s)
Adelina tribolii, *Farinocystis tribolii*	Stored product pests
Malameba locustae	Grasshoppers
Mattesia dispora, *M. trogodermae*	Lepidopterous and coleopterous pests
Nosema locustae	Grasshoppers
Nosema algerae	Anopheline mosquitoes
Nosema pyrausta	European corn borer (*Ostrinia nubilalis*)
Nosema fumiferanae	Spruce budworm (*Choristoneura fumiferana*)
Nosema gasti	Boll weevil (*Anthonomus grandis*)
Nosema melolonthae	European cockchafer (*Melolontha melolontha*)
Vairimorpha necatrix	Lepidopterous pests
Octosporea muscaedomesticae	Muscoid flies
Lambornella stegomyiae, *L. clarki*	Mosquitoes

Nosema locustae (a pathogen of various grasshoppers and crickets), *Nosema algerae* (a pathogen of anopheline mosquitoes), *Nosema pyrausta* (a pathogen of the European corn borer, *Ostrinia nubilalis*), and *Vairimorpha* (*Nosema*) *necatrix* (a pathogen of many species of caterpillars; see Figure 5.6) are currently under consideration as insect control agents.

Production

Almost all protozoa of potential value as entomopathogens must be produced in living hosts. At least 15 different protozoa have been propagated in various hosts for experimental use against pest insects. The primary hosts are generally used to produce spores, but alternate hosts can sometimes be used. For example, difficulties of producing *N. algerae* in the primary-host mosquito led to the development and use of the corn earworm, *Heliothis zea*, as an alternate host. One earworm larva, when infected with spores, produces as many spores as can be obtained from 2,000 mosquito larvae. At a proper stage of disease development, infected earworms are processed to yield purified suspensions of spores. Fresh spore suspensions have been used in most field studies, but spores usually can be stored for several months in an aqueous medium or in insect cadavers at low temperatures (0°–5°C). Spores of some species can withstand freeze-drying for longer periods of time, but some, like *N. algerae*, are inactivated upon drying.

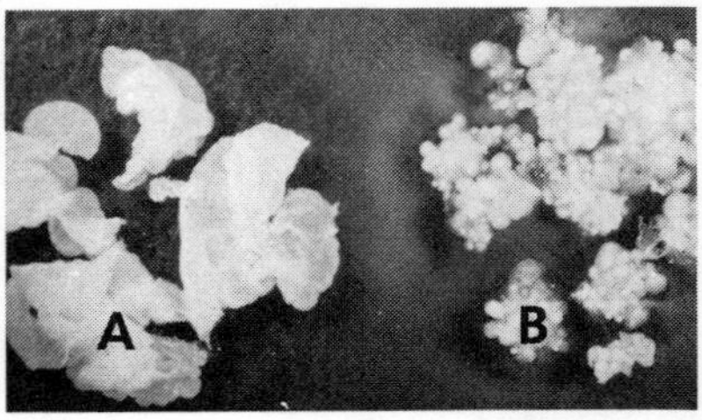

(a) Stunted and dead larvae of the corn earworm, *Heliothis zea,* infected with the microsporidium *Vairimorpha necatrix.*

(b) Comparison of nature of the fat body tissue of healthy larva (A), with that of an infected larva (B).

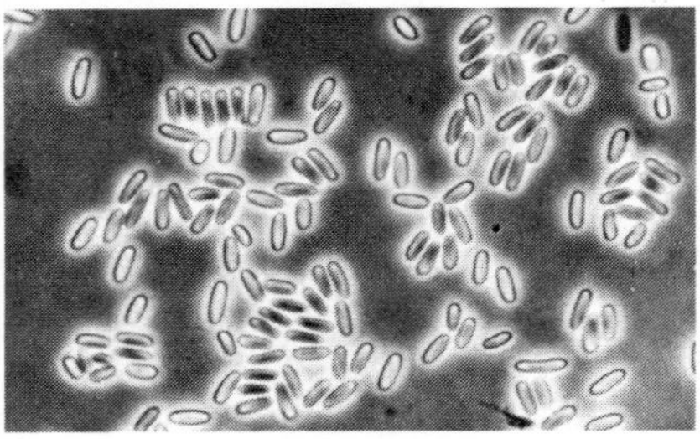

(c) Spores of *V. necatrix* from fat body of infected larva.

FIGURE 5.6 Effects of protozoa on insects. (Photographs courtesy of W. M. Brooks)

Safety and Specificity

Until recently, most protozoa were considered relatively host-specific. More extensive study of host range has revealed, however, that many are infective not only for host species in closely related genera but also for species in other families and even other orders of insects. Generally, however, although the host range for protozoa species may be quite extensive (more than 60 species of grasshoppers and crickets are susceptible to *N. locustae*), it is confined to closely related species.

Plants, human beings, and animals seem to be resistant to infection by insect protozoa. In an extensive study of *N. algerae*, a pathogen of many species of anopheline mosquitoes, no evidence of infection was found in such nontarget organisms as crayfish, freshwater shrimps, mosquito fish, several aquatic entomophagous insects, mice, or chickens.

Efficacy

Field infestations caused by protozoa have seldom been documented, but the efficacy of protozoa as control agents is readily apparent in laboratory colonies of insects. Protozoa decimate insect colonies through gradual debilitation of the host. Even in natural populations, protozoa are probably more important than we have recognized. *Nosema pyrausta* is credited with being the most important factor in maintaining corn borer populations at levels that facilitate economic control by other means. Another example is the collapse of a field population of the spruce budworm due to the debilitating effects of *Nosema fumiferanae.*

Only limited field studies of protozoa as bioinsecticides have been conducted. Most protozoa are considered more suitable for long-term suppression programs. The slower, debilitative effects of chronic infections, such as reduced fertility and shortened life span, generally preclude their widespread use as quick-acting agents.

Limitations

The availability of many kinds of protozoa enhances their potential use as microbial control agents. Although protozoa may be significant as natural regulatory agents, their use as microbial insecticides has been limited, especially by the fact that protozoa act slowly on their hosts in contrast to the rapid action of some viruses or bacteria. This means that the control of insect damage to crops during the season of treatment with a particular protozoa is usually not possible. Usually the degree of success of the treatment is directly related to timing of application.

The expense and difficulty of producing and storing sufficient quantities of protozoan spores for field use is another important limiting factor.

Large-scale field evaluations of efficacy have been undertaken only with *Nosema locustae*, a pathogen of grasshoppers. Thus, the use of protozoa as bioinsecticides is in its infancy, and much additional research will be necessary before the potential of this method can be realized.

Research Needs

Some of the more important research and development needs for the development of protozoa as bioinsecticides include:

- Establishing and maintaining a repository of viable protozoa;
- Evaluating host range and safety of promising candidates;
- Developing technology for mass production and storage; and
- Improving technology for increased field persistence and dispersal, as well as critical assessment procedures for evaluation of field efficacy.

Fungi

More than 750 fungal species representing approximately 100 genera have been reported to infect insects. Nearly all major fungal groups are represented, and virtually every type of insect is represented. Although the potential of fungi for control of insect pests in underexploited, a few isolates are being developed (Table 5.7). One of these (*Beauveria bassiana*) is already in extensive routine use in agriculture in the Soviet Union for control of the Colorado potato beetle (*Leptinotarsa decemlineata*) and the codling moth (*Laspeyresia pomonella*). A number of Russian industrial plants are producing an estimated 20 t annually of *B. bassiana* formulation for commercial use. Since little survey work has been done and few of the many known fungi have been studied extensively, it is obvious that other promising candidates can be expected in the future.

Figure 5.7 shows the steps involved in a research and development program for using a fungus to control aphids on potatoes.

Production

Means of mass production of fungi vary, but in general the available production technology is simple and straightforward. *Coelomomyces* and some *Entomophthora* species grow poorly in laboratory media, but most pathogenic

TABLE 5.7 Experimental Fungi for Insect Control

Fungus	Infective Stage	Insect Hosts	Habitat
Chytridiomycetes			
Coelomomyces	Motile planonts	Mosquitoes	Aquatic
Oomycetes			
Lagenidium	Motile zoospores	Mosquitoes	Aquatic
Zygomycetes			
Entomophthora	Conidia or resting spores	Caterpillars, aphids	Foliage
Deuteromycetes			
Aschersonia	Conidia	White flies	Foliage
Beauveria	Conidia	Beetles, caterpillars	Foliage
Hirsutella	Conidia	Mites	Foliage
Metarhizium	Conidia	Froghoppers, leafhoppers, beetles, mosquitoes	Foliage, soil, aquatic
Nomuraea	Conidia	Caterpillars	Foliage
Paecilomyces	Conidia	Beetles	Foliage
Verticillium	Conidia	Aphids, white flies	Foliage, greenhouses

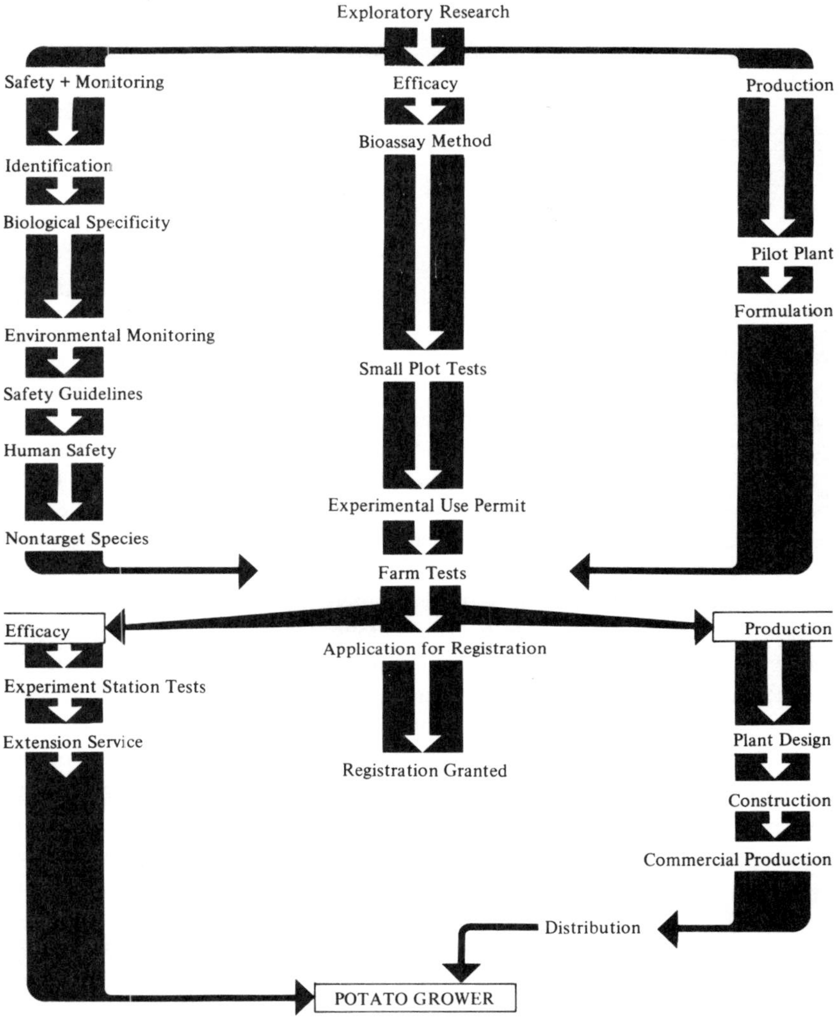

FIGURE 5.7 Development of fungal control agent for aphids. (Flow chart courtesy of R. S. Soper)

fungi, particularly species of the Fungi Imperfecti, can be mass produced on sterilized bran, grain, or beans. Semisolid fermentation or two-stage fermentation (deep fermentation for mycelial growth followed by incubation in shallow pans for spore production) have been successfully used for large-scale production of *Beauveria* and *Metarhizium*. Conidia (spores) normally are not produced in deep fermentation, but recently devised media enable production of *Beauveria* conidia. Thus, fungi can be produced in virtually any locale with technology geared to available facilities and staff.

The conidia of *Entomophthora*, because they are fragile and short-lived, are difficult to store. Many species, however, produce resting spores that survive for years. Until recently, methods for germinating these spores were unavailable, but high levels of germination now have been obtained with several species, and this opens the possibility of commercialization of these fungi.

Safety and Specificity

The host specificity of entomopathogenic fungi range from a single insect species to most members of a family or order, sometimes extending to several orders of insects and mites, and even to plants. Specificity varies not only with fungal species, but also with physiological races within species. Although *B. bassiana* occurs worldwide, isolates from one location have been shown to be more pathogenic for the same host elsewhere. The same effect has been shown for *E. sphaerosperma.* In general, fungi have been used to control a single pest, but recent work with *Nomuraea rileyi* has demonstrated control of several caterpillar pests of soybeans by conidia (Figure 5.8). With most pathogenic fungi, beneficial insects are not harmed.

With a single known exception, fungi with high virulence for insects do not cause disease in animals. Only *Conidiobolus coronatus* has shown possible infectivity to mammals. Its use as an insecticide is, therefore, not recommended. Some isolates of *Aspergillus flavus* that infect insects may also produce metabolites that are toxic to human beings, but extensive toxicological studies have not been conducted. Since pathogenic fungi usually do not grow in nature except in or on their insect hosts, toxin build up in the environment after fungi are introduced is not expected.

Efficacy

Some promising field results have been obtained with fungi. As noted, the types of insects involved in recent work are listed in Table 5.7. Fungi used in most microbial control studies were originally isolated from the pest insect or a closely related insect. In two recent studies (*Hirsutella thompsonii vs.* the citrus rust mite and *Nomuraea rileyi vs.* caterpillars), outbreaks of the fungi were artificially induced early enough to protect the crop. This procedure may be applicable to other fungus-host combinations. For example, the genus *Entomophthora* frequently causes dramatic reductions in older populations of grasshoppers, aphids, caterpillars, and flies (Figure 5.9).

Control of the codling moth (*Laspeyresia pomonella*) comparable to that obtained with chemical pesticides has been reported from Eastern Europe and Asia with *Beauveria bassiana.* Long-term effects (2–3 years) on the Colorado potato beetle (*Leptinotarsa decemlineata*) population through application of *B. bassiana* have been reported.

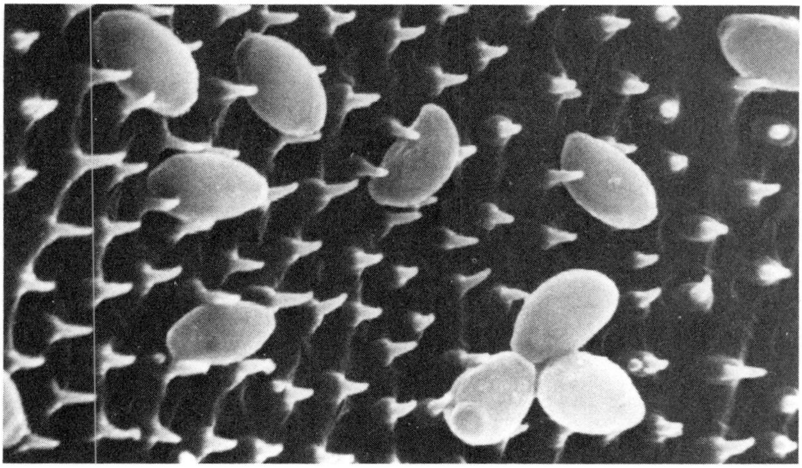

(a) Conidia of the entomopathogenic fungus *Nomuraea rileyi* on the surface of infected cotton bollworm larva (below).

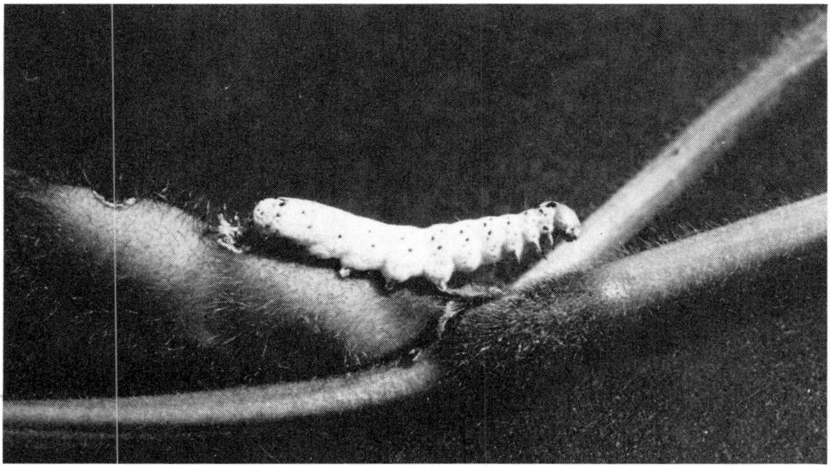

(b) Cotton bollworm larva infected with *Nomuraea rileyi.*

FIGURE 5.8 Effects of fungi on insects. (Photographs courtesy of C. M. Ignoffo)

Small-scale field tests indicate that *B. bassiana*, *B. brongniartii*, and *M. anisopliae* have good potential for control of soil-inhabiting insects such as wireworms and cockchafers, and of others such as lepidopterous larvae. The soil environment, because the humidity is usually high, is considered favorable for infection of insects by fungi. *M. anisopliae* is more frequently collected in nature from beetle grubs in the soil than from any other source, and this fungus may yield specific strains that will prove effective for control of soil insects.

FIGURE 5.9 *Sarcophaga aldrischii* infected with the fungus *Entomophthora bullata*. (Photograph courtesy of R. S. Soper)

Little is known concerning interactions between soil microorganisms and pathogenic fungi, but mortality rates can be considerably lower in unsterilized than sterilized soil. More information is needed on soil inhibitors in order to select fungal pathogens that have increased ability to infect insects in the presence of soil microorganisms.

The aquatic habitat is another promising place to use fungi. At least three fungi (*Lagenidium*, *Culicinomyces*, and *Metarhizium* species) have been developed for small-scale field trials against mosquito larvae. All have broad host ranges within the mosquito family *Culicidae*.

Limitations

Current limitations to using fungi are related to lack of knowledge rather than to negative characteristics of the fungi. Although it was generally felt that fungal control would be effective only under conditions of high temperature and humidity, it is now recognized that microclimate is more important than geographical climate, and that the microclimate can be manipulated (by irrigation, for instance, or by spacing plants to provide closed canopies).

The supposition that inoculum is already present in the environment, making introduction of fungi unnecessary, is also erroneous. Numerous field experiments have demonstrated that applications of fungal materials, particularly to foliage, greatly increase disease incidence.

A final limitation is the incomplete safety information on entomopathogenic fungi. Current data indicate that safe products can be produced, but most fungi require more extensive research before large-scale fieldwork can be conducted.

Research Needs

Several areas of research need to be explored to develop the full potential of fungal insecticides. These include:

- Conducting exploratory work to isolate new entomopathogenic fungi and improving methods for distinguishing fungal species and strains;
- Developing bioassay techniques;
- Selecting strains with increased virulence or other traits that will increase their effectiveness as fungal insecticides;
- Documenting the modes of disease induction and development;
- Developing predictive techniques so that insecticide applications can be eliminated where conditions indicate that fungi will soon significantly reduce the pest population;
- Developing methods for encouraging natural epizootics;
- Improving fungal insecticide production, formulation, and stabilization technology;
- Devising field application and evaluation methods specifically for fungal insecticides;
- Integrating fungal insecticides into pest-management systems by exploring compatibility and synergism with current chemical pesticides and cultural techniques; and
- Initiating more extensive safety tests with fungal insecticides.

References and Suggested Reading

Bacteria

Afrikian, E. G. 1973. *Entomopathogenic bacteria and their significance.* Yerevan: Armenian S.S.R. Academy of Sciences.

Bulla, L. A., Jr.; Costilow, R. N.; and Sharpe, E. S. 1978. Biology of *Bacillus popilliae. Advances in Applied Microbiology* 23:1-18.

——— ; Rhodes, R. A.; and St. Julian, G. 1978. Bacteria as insect pathogens. *Annual Review of Microbiology* 29:163-190.

Falcon, L. A. 1971. Use of bacteria for microbial control of insects. In *Microbial control of insects and mites*, H. D. Burges and N. W. Hussey, eds., pp. 67-95. New York: Academic Press.

St. Julian, G.; Bulla, L. A., Jr.; Sharpe, E. S.; and Adams, G. L. 1973. Bacteria, spirochetes, and rickettsia as insecticides. *Annals of the New York Academy of Sciences* 217:65-75.

Somerville, H. J. 1973. Microbial toxins. *Annals of the New York Academy of Sciences* 217:93-108.

Viruses

Falcon, L. A. 1976. Problems associated with the use of arthropod viruses in pest control. *Annual Review of Entomology* 21:305-324.

Ignoffo, C. M. 1973. Development of a viral insecticide: concept to commercialization. *Parasitology* 33:380-406.

———. 1973. Effects of entomopathogens on vertebrates. *Annals of the New York Academy of Sciences* 217:141-164.

McClelland, A. J., and Collins, P. 1978. UK investigates virus insecticides. *Nature* 276:548-549.

Stairs, G. R. 1971. Use of viruses for microbial control of insects. In *Microbial control of insects and mites*, H. D. Burges and N. W. Hussey, eds., pp. 97-124. New York: Academic Press.

Summers, M. D.; Engler, R.; Falcon, L. A.; and Vail, P. V. 1975. *Baculoviruses for insect pest control: safety considerations.* Washington, D.C.: American Society for Microbiology.

Summers, M. D., and Kawanishi, C. Y. 1978. *Viral pesticides: present knowledge and potential effects on public and environmental health.* Report EPA-600/9-78-026. Washington, D.C.: U.S. Environmental Protection Agency.

World Health Organization. 1973. *The use of viruses for the control of insect pests and disease vectors.* Report of Joint FAO/WHO Meeting on Insect Viruses. WHO Technical Report Series No. 531. Geneva: World Health Organization.

Protozoa

Brooks, W. M. 1974. Protozoan infections. In *Insect diseases*, G. Cantwell, ed., pp. 237-300. New York: Marcel Dekker.

McLaughlin, R. E. 1971. Use of protozoans for microbial control of insects. In *Microbial control of insects and mites*, H. D. Burges and N. W. Hussey, eds., pp. 151-172. New York: Academic Press.

Sprague, V. 1977. Systematics of the microsporidia. In *Comparative pathobiology, Vol. II: Systematics of the microsporidia*, L. A. Bulla, Jr., and T. C. Cheng, eds., pp. 1-510. New York: Plenum Publishing Corporation.

Tanada, Y. 1976. Epizootiology and microbial control. In *Comparative pathobiology, Vol. I: Biology of the microsporidia*, L. A. Bulla, Jr., and T. C. Cheng, eds., pp. 247-279. New York: Plenum Publishing Corporation.

Weiser, J. 1961. *Die mikrosporidien als parasiten der insekten.* A monograph published by *Zeitschrift fur Angewandte Entomologies*, Supplement to No. 17. Hamburg: P. Parey.

Fungi

Ferron, P. 1975. Les champignons entomopathogens: évolution des recherches au cours des dix dernières années. *SROP–Section Régionale Ouest Paléarctique* (Journal published by O.I.L.B.–Organisation Internationale de Lutte Biologique Contre les Ennemis des Cultures, Swiss Federal Institute of Technology, Zurich, Switzerland) No. 3.

McCoy, C. W. 1974. Fungal programs and their use in the microbial control of insects and mites. In *Proceedings of the summer institute on biological control of plants,*

insects and diseases, F. G. Maxwell and F. A. Harris, eds., pp. 564-575. Jackson, Mississippi: University of Mississippi Press.
Muller-Kogler, E. 1965. *Pilzkrankheiten bei insekten*. Hamburg: P. Parey.
Roberts, D. W., and Yendol, W. G. 1971. Use of fungi for microbial control of insects. In *Microbial control of insects and mites*, H. D. Burges and N. W. Hussey, eds., pp. 125-149. New York: Academic Press.

Research Contacts

Bacteria

K. Aizawa, Institute of Biological Control, Kyushu University, Fukuoka, Japan.
J. N. Aronson, Department of Chemistry, State University of New York, Albany, New York 12246, U.S.A.
L. A. Bulla, Jr., U.S. Grain Marketing Research Laboratory, U.S. Department of Agriculture, Science and Education Administration, Manhattan, Kansas 66502, U.S.A.
C. M. Ignoffo, Biological Control of Insects Research Unit, U.S. Department of Agriculture, Science and Education Administration, Federal Research, North Central Region, P. O. Box A, Columbia, Missouri 65201, U.S.A.
M. M. Lecadet, Institut de Recherches en Biologie Moléculaire, Centre National de la Recherche Scientifique, Université de Paris VII, 2 place Jussieu, 75221, Paris, France.
A. A. Yousten, Department of Biology, Virginia Polytechnic Institute and State University, Blacksburg, Virginia 24061, U.S.A.

Viruses

H. C. Chapman, Gulf Coast Mosquito Research Laboratory, U.S. Department of Agriculture, Science and Education Administration, Lake Charles, Louisiana 70601, U.S.A.
L. A. Falcon, Department of Entomology and Parasitology, 333 Hilgard Hall, University of California, Berkeley, California 94720, U.S.A.
C. M. Ignoffo, Biological Control of Insects Research Unit, U.S. Department of Agriculture, Science and Education Administration, Federal Research, North Central Region, P. O. Box A, Columbia, Missouri 65201, U.S.A.
G. R. Stairs, Department of Entomology, The Ohio State University, Columbus, Ohio 43210, U.S.A.
C. Vago, Station de Recherches de Pathologie Comparée INRA-CNRS-EPHE, Université des Sciences, Place Eugene Bataillon, 34060 Montpellier, France.

Protozoa

W. M. Brooks, Department of Entomology, North Carolina State University, Raleigh, North Carolina 27650, U.S.A.
E. I. Hazard, Insects Affecting Man Research Laboratory, U.S. Department of Agriculture, Science and Education Administration, Gainesville, Florida 32604, U.S.A.
J. E. Henry, Rangeland Insect Laboratory, U.S. Department of Agriculture, Science and Education Administration, Montana State University, Bozeman, Montana 59715, U.S.A.
J. V. Maddox, Section of Economic Entomology, Illinois Natural History Survey, Urbana, Illinois 61801, U.S.A.
J. Weiser, Laboratory of Insect Pathology, Institute of Entomology, Academy of Sciences, Flemingovo nam 2, Praha 6, Czechoslovakia.

Fungi

J. P. Latge, Parasitologie Végétale, Institut Pasteur, 25 Rue du Dr. Roux, Paris, France.
C. W. McCoy, Department of Entomology, University of Florida, Lake Alfred, Florida 33850, U.S.A.

D. W. Roberts, Boyce Thompson Institute for Plant Research, Tower Road, Ithaca, New York 14853, U.S.A.

R. S. Soper, U.S. Department of Agriculture, Science and Education Administration, Insect Pathology Research Unit, Boyce Thompson Institute, Cornell University, Tower Road, Ithaca, New York 14853, U.S.A.

D. Tyrrell, Great Lakes Forest Research Centre, P. O. Box 490, Sault Ste. Marie, Ontario P6A 5M7, Canada.

J. Weiser, Laboratory of Insect Pathology, Institute of Entomology, Academy of Sciences, Flemingovo nam 2, Praha 6, Czechoslovakia.

N. Wilding, Rothamsted Experiment Station, Harpenden, Herts. AL5 2JQ, England.

Chapter 6

Fuel and Energy

Microorganisms have a historic role in the fermentation of a variety of organic materials to alcohols, acids, and CO_2, mainly related to the manufacture of foods, beers, and wines. Today alcohol production is largely synthetic, that is, nonmicrobial, although the rising costs of petroleum have created renewed interest in the production of ethanol by fermentation for use as a fuel. For example, ambitious programs for production of large quantities of fuel ethanol by fermentation have been undertaken in Brazil, India, and several other countries. Similarly, biogas is being produced as a source of energy in several countries.

The most practical process to produce fuel for farm and community use by microbial processes is the generation of biogas (Table 6.1). Ethanol for fuel requires a capital investment more in keeping with regional or large farm-cooperative manufacture. The technology for both methane and ethanol manufacture is readily accessible.

The generation of hydrogen and methanol through microbial processes is still in the laboratory stage. Although there are organisms that yield hydrogen and methanol from organic substrates, much more development work is required to make these processes economically feasible.

TABLE 6.1 Characteristics of Fuels from Microbial Processes

Fuel	Approximate Gross Energy Content	Typical Sources	Process Considerations
Liquids			
Methanol	10,000 BTU/lb 23.8 MJ/kg	Methane	None commercially available
Ethanol	13,000 BTU/lb 30.6 MJ/kg	Molasses, grains, plant biomass	Requires significant capital investment
Gases			
Methane	24,000 BTU/lb 55.5 MJ/kg	Animal, human, and agricultural wastes	Practical for farm and community use
Hydrogen	61,000 BTU/lb 142 MJ/kg	Algae-nutrient system	None commercially available

Ethanol

The production of ethanol from residues with a high sugar content may soon be economically practical as a means of offsetting the rising costs of petroleum. Ethanol can be used alone or blended with gasoline or diesel fuels. For this use it need not be of high purity or entirely free of water. The alcohol yield depends on the amount of starch or fermentable sugars present in the substrate; sugar-cane is a suitable raw material because of the large amounts of this crop available in most parts of the world.

The economics of ethanol production improve as the size of the plant increases, to the point where costs of collecting raw material to sustain a huge processing unit become too large. For a plant with a capacity of 20 million gallons per year, capital investment would be about $30 million. Plant schematics and material requirements for the conversion of molasses and corn to ethanol are shown in Figures 6.1 and 6.2. Estimates of the areas required for various crops to support a 100,000 t per year plant are shown in Table 6.2.

The fermentations shown in Figures 6.1 and 6.2 are conducted at atmospheric pressure. However, for the production of volatile products like ethanol, both rapid fermentations using a vacuum, and the recycling of microbial cells have significant advantages over older conventional methods. When a vacuum of approximately 50 mm of mercury is applied to a fermenter operating at 35°C, the ethanol can be removed continuously as it is produced by yeast. The removal of ethanol overcomes inhibition of the fermentation. Reductions in yield and productivity and suppression of yeast growth occur at ethanol concentrations of 7–10 percent. High substrate concentrations can be used in vacuum fermentations and good yields still

TABLE 6.2 Crop Area to Support a 100,000 T Per Year Ethanol Plant

	Africa	(hectares, in thousands) South America	Near East	Far East
Corn	349	253	154	333
Wheat	457	268	370	305
Rice	213	213	76	157
Cassava	87	49	135	73
Sugar Cane:				
Molasses only	120	125	78	145
Total Cane Juice	36	37	23	43

Source: Leo Hepner. 1977. Feasibility of producing basic chemicals by fermentation. In *Microbial energy conversion*, H. G. Schlegel and J. Barnea, eds. Oxford: Pergamon Press. p. 550.

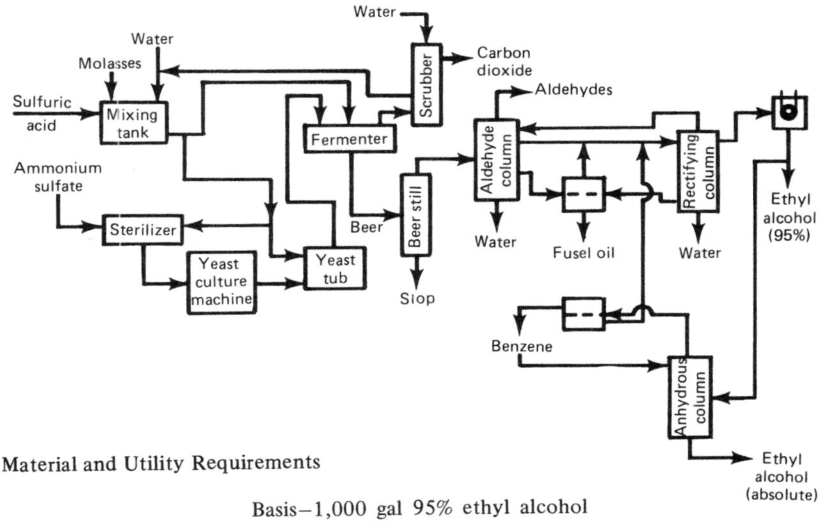

Material and Utility Requirements

Basis–1,000 gal 95% ethyl alcohol

plus 4 gal fusel oil, 4,800 lb carbon dioxide, 1,000 lb carbon, and 900 lb potash

Molasses (blackstrap)	2,400 gal	Process water	10,000 gal
Sulfuric acid (60° Bé)	170 lb	Cooling water	42,000 gal
Ammonium sulfate	15 lb	Electricity	110 kwhr
Steam	50,000 lb		

Source: W.L. Faith; D. B. Keyes; and R. L. Clark, 1974. *Industrial chemicals.* New York: John Wiley and Sons.

FIGURE 6.1 Ethanol from molasses by fermentation.

achieved. When both vacuum fermentation and cell recycling are practiced, productivities from 10 to 12 times higher than conventional batch processes are achieved. Increases in productivity reduce capital costs and energy requirements for fermenter operation.

Vacuum distillate from the fermentation liquid contains up to 20 percent ethanol. The cost of distillation to achieve an ethanol concentration of 80–95 percent is then significantly less than that of conventional methods. Application of these techniques for the production of ethanol should make it more attractive as a substitute for fossil fuel.

Production of ethanol and methane from cellulosic wastes by thermophilic microbes is promising for future development. Various lignocellulose sources including rice straw, corn stover, bagasse, etc., may be the ultimate choices for feedstocks because of their lack of alternative value as foods.

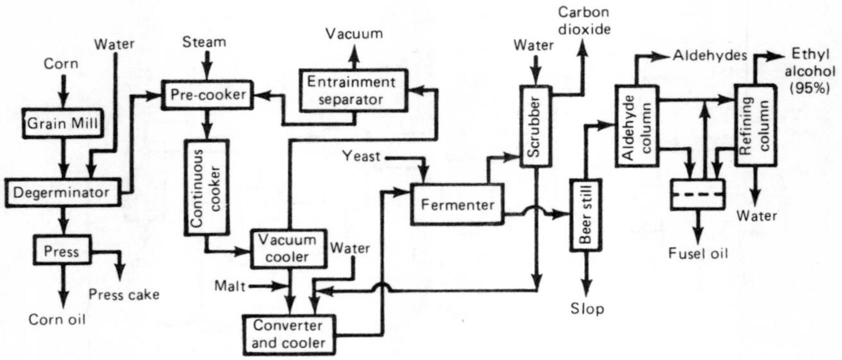

Material Requirements

Basis—1,000 gal 95% ethyl alcohol

plus 400 lb corn oil, 10 gal fusel oil, 4,800 lb carbon dioxide, 750 lb press cake, and 4,400 lb stock feed (dry residue)

Corn	372 bu	Yeast	Variable
Barley malt	83 bu	Process water	17,000 gal

Source: W. L. Faith; D. B. Keyes; and R. L. Clark, 1974. *Industrial chemicals.* New York: John Wiley and Sons.

FIGURE 6.2 Ethanol from corn to fermentation.

Limitations

Village production of fuel-grade ethanol (95 percent and above) is inappropriate because there are no economical small-scale concentration techniques available. Large-scale production requires harvest and transport of the crops over a fairly broad area, and costs for these operations must be included.

Research Needs

The following research and development activities are needed to facilitate widespread microbial production of ethanol:

- A survey of the types and characteristics of raw materials for fermentation in various climates and locales;
- Agricultural research to improve yields of crops, such as cassava and sorghum, which are candidate substrates for alcohol fermentations;
- Improvements in methods for harvesting and preparation of fermentable substrates from a variety of agricultural, forestry, and other organic wastes and crops;

• Development of equipment and processes for fermentation and recovery of alcohol that require less capital, energy, and labor than conventional processes; and

• Socioeconomic studies to develop optimal systems for harvest, transport, and processing of crops.

Utilization of Cellulose

Several anaerobic bacteria of the genus *Clostridium* have been used in fermentations of cellulose. *C. thermocellum*, which has simple nutritional requirements, is the only known thermophilic species that degrades cellulose. Because it grows at higher temperatures (above 50°C) than most bacteria, it has the advantage of being less prone to contamination and also has a faster reaction rate than microbes growing at lower temperatures. In pure culture fermentations, the chief products from cellulose are cell mass, acetate, ethanol, lactate, hydrogen, and carbon dioxide. In a mixed culture of *C. thermocellum* and *Methanobacterium thermoautotrophicum*, the major products from cellulose are cell mass, methane, and acetate. One can envision the use of *C. thermocellum* in pure cultures for ethanol production or in mixed cultures for the production of biogas from cellulosic wastes.

Limitations

The accumulation of acetic acid during fermentation limits growth, and since *C. thermocellum* cannot decompose lignin, many natural substrates such as wood must be pretreated by acid hydrolysis to make the cellulose available for fermentation.

Research Needs

The following requirements are necessary to facilitate the utilization of cellulose as an energy source:

• Process development for specific substrates and end-products as well as for an optimal fermenter design;

• Better understanding of the biochemistry of the process; and

• Development of a means for converting the acetate to nontoxic products.

Methane

Microbiological conversion of organic materials to methane (biogas) is a natural process, providing energy in a clean, gaseous form. Although this process will not meet the total energy demands of modern society, it may economically supplement other sources of fuel. Its use will depend on factors such as the cost of fossil fuels, availability and degradability of organic substrates, and the availability of trained personnel.

Biogas production occurs in many natural microbial ecosystems such as organic sediments of aquatic systems, marshes and soil, and in the rumen and large intestine, especially in herbivorous animals. It involves a complex mixture of anaerobic bacteria, which convert up to 90 percent of the combustible energy of the degradable organic matter to methane and carbon dioxide.

Anaerobic treatment of complex organic mixtures may be considered a three-stage process, as shown in Figure 6.3. In the first stage, a group of facultative microorganisms acts upon the organic substrates. By enzymatic hydrolysis, the complex substances are solubilized and serve as the substrates for microorganisms in the second stage. In the second stage, these soluble organic compounds are converted to organic acids. The acids (primarily acetic) serve as substrates for the final stage of decomposition accomplished by the methanogenic bacteria. These bacteria can produce methane either by converting acetic acid to methane and carbon dioxide or by reducing carbon dioxide to methane, using hydrogen or formate produced by other bacteria. Only about 10 percent of the energy is converted into microbial cells that obtain energy for growth during the conversion.

With organic waste materials such as cattle manure or urban organic refuse, in theory 30-50 percent of the combustible energy could be converted to methane. With an efficiently operated digester and substrate materials such as cattle waste, as much as 4.5 liters of methane can be produced per liter of reactor material each day. With some vegetable materials or forages, even higher rates are possible, and as much as 70 percent of the energy can be converted to methane.

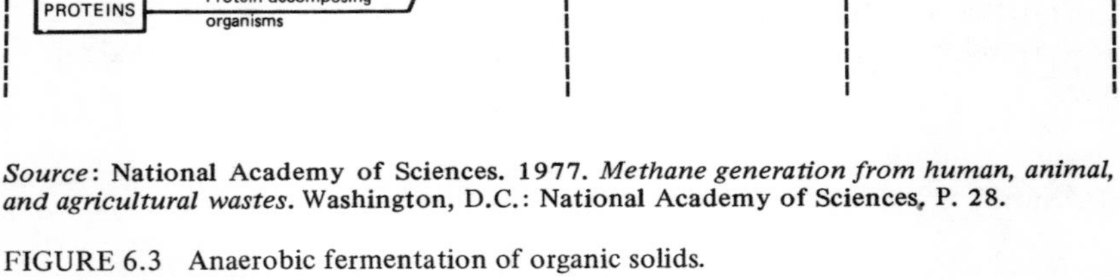

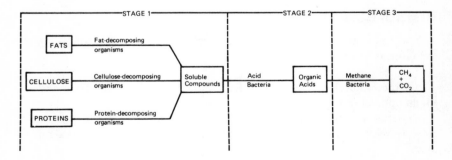

Source: National Academy of Sciences. 1977. *Methane generation from human, animal, and agricultural wastes*. Washington, D.C.: National Academy of Sciences, P. 28.

FIGURE 6.3 Anaerobic fermentation of organic solids.

The economically valuable substances such as ammonia nitrogen, phosphate, and microbial cells are retained in the reactor effluent and residue, and these may have value as fertilizer or as an animal feed supplement.

The residue also has value as a soil conditioner, and it usually does not attract insects or have the disagreeable odors often associated with animal manure. Such residues can also be dried and burned to obtain additional heat energy. The fermentation process can be applied to sewage effluent as a step in water recycling by removing nutrients and facilitating subsequent water purification (see Chapter 8). The gas produced can be an economic incentive for waste treatment. In addition, the process also converts malodorous and pathogenic waste into an innocuous, potentially useful sludge.

The technology of biogas production is highly developed and can be applied economically to many organic substrates, depending on their biodegradability, their alternative uses, the possible economic value of by-products, and the competitive use of fossil fuel. Thousands of small-scale plants of family, farm, or village size have been operated, especially in parts of Asia and in Europe.

A typical methane plant in India based on cow manure is shown in Figure 6.4.

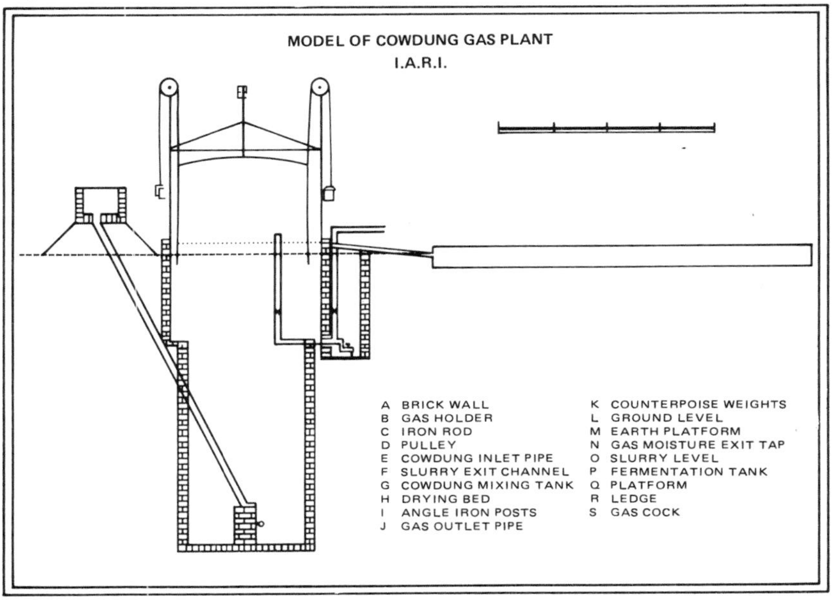

Source: National Academy of Sciences. 1977. *Methane generation from human, animal, and agricultural wastes.* Washington, D.C.: National Academy of Sciences. P. 70.

FIGURE 6.4 Biogas plant designed by Acharya, developed at Indian Agricultural Research Institute.

Methane can be separated from other contaminating gases, such as carbon dioxide and hydrogen sulfide, and burned to generate electricity and heat and to produce steam, or for cooking. Biogas can be added as a supplement to natural gas pipelines, stored in tanks under atmospheric pressure, or compressed for storage.

Capital and maintenance costs of biogas production units vary with the size of the plant, which is a function of local resources and needs. Table 6.3 shows some cost estimates for various sizes of methane plants in India.

TABLE 6.3 Capacities and Costs of Biogas Plants

Daily Production Rate (m^3)	Estimated Cost in 1975 (in Rupees)	Approximate Head of Cattle or Swine Required
3	3,016	3–4
6	4,175	6–10
10	6,100	16–20
20	11,500	35–40
45	20,740	60–70
85	38,800	110–140
140	58,000	400–450

Source: National Academy of Sciences. 1977. *Methane generation from human, animal, and agricultural wastes.* Washington, D.C.: National Academy of Sciences. P. 120.

Limitations

Temperature is a very important factor in biogas production. The rate of fermentation decreases rapidly below 20°C. Maximum rates are obtained at about 40°C (optimum for many mesophilic bacteria), but for thermophilic bacteria the optimum temperature is near 60°C. The most effective temperature for operation of a given digester depends on a number of factors such as insulation, climate, rate of conversion of the substrate, and concentrations of nutrients and other substances formed from the breakdown of the substrate.

If dissolved materials in the effluent from the fermentation are low and the substrate being used is high in dry matter, it is possible to recycle the effluent and thereby conserve heat and nutrients. If nutrient concentrations are high (as in animal wastes), recycling of effluent liquids is not practical, and effluent heat value can be conserved only through heat exchangers.

The system must be maintained in the neutral range (pH 6.8–7.8) for optimum rates. The pH may vary according to concentrations of acids or ammonia in the reactor. The percentage of solids and the biodegradability of substrates in the feed are important. With a high proportion of solids, the energy requirements for mixing, which is essential for effective fermentation, may be too high and soluble substances such as ammonia or organic acids may reach toxic concentrations. On the other hand, dilute substrates require large digesters. Nutrient elements such as nitrogen, phosphorus, sulfur, and iron may need to be supplied in an appropriate form and concentration; many materials such as human and animal wastes (but not urban refuse) contain adequate quantities of these substances.

The reaction time, or rate of substrate use measured as flow per volume of reactor per day, needs to be short for rapid methane production. But if the reaction time is too short, less organic material is degraded and less methane is produced. Bacteria are sensitive to abrupt changes in loading rate and retention time, and they lose the potential for rapid activity during long periods of inactivity when substrate is not added. An efficient, rapid methane-producing system requires continuous or semicontinuous feeding, without drastic changes in operating conditions. Because of heating and other demands, larger digesters are generally more economical.

The suitability of organic substrates for methane production depends on many factors. For example, removal or addition of water from or to relatively wet or dry substrates may be expensive. The inherently higher water content of some substrates will increase transportation costs and require a larger reactor. Collection costs and incentives for collection may also be important.

Organic sugars fermented anaerobically produce roughly equal volumes of methane and carbon dioxide. The CO_2 reduces the energy content of the biogas. It may be removed by "scrubbing" it by passing it through dilute alkali. Many organic materials such as cattle wastes, forages, and urban wastes contain large amounts of lignin, silicates, and waxes that are indigestible under anaerobic conditions. These substances may greatly reduce the extent of conversion of cellulosic materials to methane. Treatment to increase the fermentability of such materials may be expensive. Some substrates, such as urban refuse, require separation of nonorganic wastes like glass and metals, and some may require particle-size reduction for effective fermentation. Many effective substrates, including whole corn, alfalfa, sugarcane, and cassava, may be more economically used as food or animal feed.

Residual sludge disposal may be a problem, and aqueous effluent may cause pollution. But the residue and effluent may also serve as substrates for the growth of algae, which in turn may be an effective substrate for the production of methane gas.

Even at the farm or village level, the effective production of methane requires competent management, but this should not be a problem if adequate teaching and extension activities are available.

Research Needs

The following research and development efforts are needed to facilitate greater production and use of biogas:

- Research leading to effective, economically produced substrates with emphasis on those that are photosynthetically derived;

- Development of more economical collection and processing procedures—for example, by dewatering or desalting—along with economical means of increasing fermentability;
- Research on separation of the cellulose-lignin complex;
- Identification of economic uses for lignin and other residues;
- Investigation of the relationship of the chemical composition of substrates to their efficiency in conversion to methane; and
- Design of improved and less-expensive digesters or reactors and their components, for instance, solar energy and better heat exchangers, to reduce the energy required to maintain temperatures at which fermentation is most effective.

Methanol

Methane gas is produced in enormous amounts in some areas of the world as a by-product of oil recovery and refining. It is also found in mud at the bottom of marine and freshwater environments and as a product of anaerobic fermentation of organic wastes.

Methane is often difficult to transport from areas where it is produced to parts of the world where it can be used as an energy source. It must either be distributed by pipelines or refrigerated at −162°C for transport by tanker. Since tanker transport is too expensive and hazardous to justify in most cases, the gas is simply flared off in many oil fields.

By contrast, methanol has a relatively high energy content per unit volume, and its transport is less expensive. Methanol is used in many parts of the world for heating, lighting, cooking, and power. It also has potential as a nonpolluting industrial fuel.

The conversion of methane to methanol as an intermediate in the oxidation of methane to carbon dioxide occurs in all methane-oxidizing bacteria. Methanol accumulates in small amounts in many pure culture fermentations. The use of inhibitors (iodoacetate, phosphate buffers, and EDTA) of the microbial enzyme methanol dehydrogenase has been demonstrated. These materials inhibit methanol utilization without preventing the preceding methane oxidation step. In addition, some work has been conducted on "leaky" mutants of methane oxidizers that excrete (rather than further oxidize) methanol. Yields are poor, however, even in the best cases. The bioconversion is believed to involve the following reaction in all methane-oxidizing bacteria:

$$\underset{\text{Methane}}{CH_4} + O_2 + XH_2 \rightarrow \underset{\text{Methanol}}{CH_3OH} + H_2O + X$$

where XH_2 is an agent that reduces one-half of the O_2 molecule in the biochemical process. Thus, a reducing agent must be available, which may be obtained by concurrent oxidation of inexpensive substrates or, ideally, by biophotolysis of water.

Limitations

The biochemistry of methane oxidation was poorly understood until recently, and it is necessary to establish the energy balance of the system before the process can be evaluated.

The methanol produced will necessarily be in an aqueous solution, and the economics of recovery by distillation may preclude its use as a fuel unless cheaper means are found. It may be more practical to use the methanol as a fermentation substrate for single-cell protein production than to distill it to obtain a fuel.

Research Needs

When improving the feasibility of methanol production by microbial processes, the following preliminary steps should be taken:

- Elucidation of the biochemical pathways of the processes;
- Identification of new microorganisms and mutants that have greater potential for producing methanol; and
- Exploration of the use of inhibitors of methanol oxidation.

Hydrogen

The enormous amounts of solar radiation that reach the earth's surface greatly exceed the world's foreseeable needs for energy. The use of solar energy to produce fuels as well as biomass directly is, therefore, an inviting technical challenge. All green plants and algae use solar radiation in the reduction of CO_2 by water. The oxygen is released as molecular oxygen and the concurrent reducing equivalents (hydrogen ions and electrons) are used to reduce carbon dioxide in order to synthesize cellular constituents. It is possible under certain conditions to modify this biological process and cause the production of hydrogen from the biologically produced reducing equivalents.

Biophotolysis, the production of hydrogen from water using the radiant energy of sunlight, has been demonstrated in a large number of algal cultures. It is theoretically possible to produce hydrogen from water using any plant or algae that contains the hydrogenase enzyme.

The use of algae as a means of trapping solar energy to expand fuel supplies is attractive for reasons other than its potential as a cheap energy alternative. Hydrogen as a fuel is nonpolluting. The substrate, water, and the energy source, sunlight, are inexhaustible.

Many algae have simple nutritional requirements and can be cultivated on dilute waste materials. They can potentially be utilized as a food or fertilizer after use as a catalyst for hydrogen production. Algae as catalysts for the process are easily renewed and can perhaps be preserved in an active state.

The most important considerations for future energy processes using algae include: 1) identification of strains that produce molecular hydrogen at the highest rates and use radiant energy most efficiently, and 2) genetic manipulation of these organisms and alteration of their metabolic processes to increase hydrogen production. A partial list of algae capable of evolving hydrogen is shown in Table 6.4.

TABLE 6.4 Algae Capable of Evolving Hydrogen

Scenedesmus obliguus	*Chlamydomonas reinhardii*
Scenedesmus quadricauda	*Ankistrodesmus brauni*
Chlorella vulgaris	*Ankistrodesmus stipitatum*
Chlorella fusca	*Dunaliella* sp.
Chlorella autotrophica	*Chondrus crispus*
Chlamydomonas moewusii	*Corallina officinals*
Chlamydomonas debaryana	*Ceramium rubrum*
Chlamydomonas dysosmos	*Porphyridium aerugineum*
Chlamydomonas humicola	

Hydrogen production has been demonstrated from cell fractions of algae supplemented with essential enzymes. The major theoretical advantages of using this approach, which requires purification of all the fractions necessary for biophotolysis, would be a reduced system size for trapping light energy and improved efficiency. Isolated cell fractions, on the other hand, are renewed with difficulty and present significant technical problems in storage and preservation of catalytic activity.

The production of hydrogen by fermentation of organic substrates using nonphotosynthetic microorganisms has been discussed as a possible process for energy production. Hydrogen is also an end-product of organic fermentation by many anaerobic bacteria.

This process as an alternative to methane production has little value for energy conservation. The maximum amount of energy that can be conserved in the hydrogen produced by fermentation is 33 percent of the energy available in the best substrates. The fermentation requires stringent precautions to ensure culture stability. By contrast, the energy conserved in the methane fermentation can exceed 80 percent of the energy available in organic matter, and this fermentation process has much greater potential for efficiency and economy.

Limitations

Biophotolysis for the production of hydrogen requires large reactors that are transparent to radiant energy and impermeable to hydrogen. The cost of any reactor must allow the process to be competitive with alternative energy sources. The production of hydrogen by most organisms is sensitive to oxygen, which must be rigorously excluded.

The algae must be grown under adequate nutritional conditions, and the environment must be modified so that biomass production ceases and hydrogen production proceeds. Hydrogen evolution can be accomplished by removal of carbon dioxide or nitrogen in some cultures. The optimum conditions will vary with the organism selected and other factors.

Research Needs

The present state of the technology for biophotolysis precludes its application in the near future. To further its potential, the following research should be undertaken:

- Examination of many kinds of algae for their potential as catalysts for hydrogen production;
- Determination of optimum conditions for efficient hydrogen production; and
- Research on microencapsulation and stabilization of subcellular fractions capable of hydrogen evolution.

Bacterial Leaching

Thiobacillus ferrooxidans is a bacterium that lives in acid environments and obtains energy for growth by oxidizing reduced (ferrous) iron in various metal sulfides, sulfur, and soluble sulfur compounds. Many insoluble metal sulfides can be oxidized to corresponding metal sulfates. The oxidation of elements in pyrite ore can lead to the production of oxidized iron, sulfuric acid, and metal salts of sulfuric acid. The oxidized iron and sulfuric acid produced by the bacteria can be used for the extraction of uranium and other metals from raw ores.

Large-scale leaching of uranium ores is employed in India, Canada, and the Soviet Union. In India, the leach liquor is percolated through a descending system of terraces containing pyritic uranium oxides. Exploitation of bacterial leaching enables the recovery of uranium from low-grade ore (0.01–0.05 percent U_3O_8), uneconomical to process by other means. The process can also be applied to high-grade material such as uranium-rich pillars supporting the roof of a worked-out mine. Recovery of uranium is accomplished by employing resin columns, and the oxidized iron in the liquid is recycled through the ore or slag.

The cost of recovering uranium from low-grade ores by leaching is less than by conventional processes. In some cases, improved methods have been developed involving the production of acid ferric sulfate by a bacterial oxidation of pyrite, with the acid ferric sulfate used to leach ground ore. Continuous-culture methods for uranium leaching have also been described. The process with the greatest commercial potential is one that requires production of a ferric iron leach liquid by a bacterial process and then utilizes the liquor for chemical leaching of uranium or other metals.

The control of the bacterial populations is relatively easy because the acid content of the leach liquor and the substrates available for growth limit the kinds of organisms that can grow.

These are practical, proven processes for recovering uranium that remain underexploited in many parts of the world. The attractions of leaching over conventional methods include its simplicity of operation, the lower capital requirements for materials, and lower energy costs.

Bacterial leaching has also been applied to the recovery of hydrocarbons from oil shale.

Research Needs

- Considerable pilot-scale research has been devoted to defining the ideal conditions for maximum rates of extraction of uranium and other metals. More attention needs to be given to scale-up of the extraction process.
- Leaching underground should be developed to obviate bringing ore to the surface.

References and Suggested Reading

Ethanol

Anderson, Earl V. 1978. Gasohol: energy mountain or molehill? *Chemical and Engineering News* 56:8-12, 15.

Cysewski, G. R., and Wilke, C. R. 1977. Rapid ethanol fermentation using vacuum and cell recycle. *Biotechnology and Bioengineering* 19:1125-1143.

Faith, W. L.; Keys, D. B.; and Clark, R. L. 1974. *Industrial chemicals.* New York: John Wiley and Sons, Inc.

Gall, Norman. 1978. *Noah's ark: energy from biomass in Brazil.* Report No. 30. Hanover, New Hampshire: American Universities Field Staff.

Hepner, Leo. 1977. Feasibility of producing basic chemicals by fermentation. In *Microbial energy conversion*, H. G. Schlegel and J. Barnea, eds., pp. 531-553. Oxford: Pergamon Press.

Jackson, E. A. 1976. Brazil's national alcohol program. *Process Biochemistry* 11:29-30.
Paturau, J. M. 1969. *By-products of the cane sugar industry.* New York: Elsevier-North Holland Publishing Company.
Rao, M. R. K., and Murthy, N. S. 1963. Alcohol as a fuel for diesel engines. Paper Presented at the Symposium on New Developments of Chemical Industries Relating to Ethyl Alcohol, Its By-products and Wastes, 14-16 October, at New Delhi.

Utilization of Cellulose

Cooney, C. L., and Wise, D. L. 1975. Thermophilic anaerobic digestion of solid waste for fuel gas production. *Biotechnology and Bioengineering* 17:1119-1135.
Ng, T. K.; Weimer, P. J. P.; and Zeikus, J. G. 1977. Cellulolytic and physiological properties of *Clostridium thermocellum. Archives of Microbiology* 114:1-7.
Weimer, P. J. P., and Zeikus, J. G. 1977. Fermentation of cellulose and cellobiose by *Clostridium thermocellum* in the absence and presence of *Methanobacterium thermoautotrophicum. Applied and Environmental Microbiology* 33:289-297.

Methane

Bryant, M. P. 1979. Microbial methane production–theoretical aspects. *Journal of Animal Science* 48:1.
Golueke, C. G. 1977. *Biological reclamation of solid wastes.* Emmaus, Pennsylvania: Rodale Press, Inc.
Gould, R. F., ed. 1971. *Anaerobic biological treatment processes.* Advances in Chemistry Series, No. 105. Washington, D.C.: American Chemical Society.
Jewell, W. J.; Davis, H. R.; Gunkel, W. W.; Lathwell, D. J.; Martin, J. A., Jr.; McCarty, T. R.; Morris, G. R.; Price, D. R.; and Williams, D. W. 1976. *Bioconversion of agricultural wastes for pollution control and energy conversion.* Final Report TID 27164, for the U.S. Department of Energy under the National Science Foundation Contract No. ERDA-NSF-741222 A01. Ithaca, New York: Cornell University, Division of Solar Energy.
National Academy of Sciences. 1977. *Methane generation from human, animal, and agricultural wastes.* Report of an *Ad Hoc* Panel of the Advisory Committee on Technology Innovation, Board on Science and Technology for International Development, Commission on International Relations. Washington, D.C.: National Academy of Sciences.
Pfeffer, J. T., and Liebman, J. C. 1976. Energy from refuse by bio-conversion, fermentation and residue disposal processes. *Resource Recovery and Conversion* 1:295.
Schlegel, H. G., and Barnea, J., eds. 1976. *Microbial energy conversion: Report of the United Nations Institute for Training and Research.* Oxford: Pergamon Press.
Van Soest, Peter J., and Mertens, D. P. 1974. Composition and nutritive characteristics of low quality cellulosic wastes. *Federation Proceedings* (published by Federation of American Societies for Experimental Biology) 33:1942-1944.

Methanol

Anthony, C. 1975. The biochemistry of methylotrophic microorganisms. *Science Progress* (London) 62:167-206.
Foo, E. L., and Hedén, C.-G. 1977. Is biocatalytic production of methanol a practical proposition? In *Microbial energy conversion*, H. G. Schlegel and J. Barnea, eds., pp. 267-280. Oxford: Pergamon Press.
Hedén, C.-G. 1974. Microbial aspects of the methanol economy. *Annual Review of Microbiology* 24:137-150.
Ribbons, D. W.; Harrison, J. E.; and Wadsinski, A. M. 1970. Metabolism of single carbon compounds. *Annual Review of Microbiology* 24:135-158.
Whittenbury, R. J.; Dalton, E. J.; and Reed, H. L. 1975. The different types of methane oxidizing bacteria and some of their more unusual properties. In *Microbial growth on C-compounds*, pp. 1-9. Kyoto, Japan: The Society of Fermentation Technology.

Hydrogen

Gaffron, H., and Rubin, J. 1942. Fermentative and photochemical production of hydrogen in algae. *Journal of General Physiology* 26:219-240.

Kondratieva, E. N. 1976. Phototrophic microorganisms as a source of hydrogenase formation. In *Microbial energy conversion*, H. G. Schlegel and J. Barnea, eds., pp. 205-216. Oxford: Pergamon Press.

Oschepko, V. P., and Krawnovski, A. A. 1976. Photoproduction of molecular hydrogen by green algae. *Akademiya Nauk S.S.R. Izvestiya. Seriya Biologicheskaya* 87:100.

Rao, K. K.; Rosa, L.; and Hall, D. O. 1976. Prolonged production of hydrogen gas by a chloroplast biocatalytic system. *Biochemical Biophysical Research Communications* 68:21-27.

San Pietro, A. 1977. Hydrogen formation from water by photosynthesis and artificial systems. In *Microbial energy conversion*, H. G. Schlegel and J. Barnea, eds., pp. 217-233. Oxford: Pergamon Press.

Stuart, T. S., and Gaffron, H. 1972. The mechanism of hydrogen production by several algae. *Planta* (An International Journal of Plant Biology) 106:101-112.

Thauer, R. K.; Jungermann, K.; and Deker, K. 1977. Energy conservation in chemotrophic anaerobic bacteria. *Bacteriological Review* 41:100-180.

Bacterial Leaching

Guay, P.; Silver, M.; and Torma, A. E. 1977. Ferrous iron oxidation and uranium extraction by *Thiobacillus ferrooxidans. Biotechnology and Bioengineering* 19:727-740.

Kelly, D. P. 1977. Extractions of metals from ores by bacterial leaching. In *Microbial energy conversion*, H. G. Schlegel and J. Barnea, eds., pp. 329-338. Oxford: Pergamon Press.

Meyer, W. Graig, and Yen, T. F. 1976. Enhanced dissolution of oil shale by bioleaching and thiobaciili. *Applied Environmental Microbiology* 32:610-613.

Touvinen, O. H., and Kelly, D. P. 1974. Use of microorganisms for the recovery of metals. *International Metallurgical Review* 19:21-31.

Warren Spring Laboratory. 1975. *A bacterially assisted process for uranium extraction.* Leaflet No. IME/75F/10c. Stevenage, Hertfordshire (England): Warren Spring Laboratory.

Research Contacts

Ethanol

W. D. Bellamy. Department of Food Science, Cornell University, Ithaca, New York 14853, U.S.A.

C. R. Wilke, Department of Chemical Engineering, University of California, Berkeley, California 94720, U.S.A.

Utilization of Cellulose

Charles L. Cooney, Department of Bioengineering, Massachusetts Institute of Technology, Cambridge, Massachusetts 02139, U.S.A.

J. G. Zeikus, Department of Bacteriology, University of Wisconsin, Madison, Wisconsin 53706, U.S.A.

Methane

Jerome Goldstein, Editor, *Compost Science/Land Utilization*, Box 351, Emmaus, Pennsylvania 18049, U.S.A.

W. J. Jewell, Department of Agricultural Chemistry, Cornell University, Ithaca, New York 14850, U.S.A.

P. L. McCarty, Department of Civil Engineering, Stanford University, Stanford, California 94305, U.S.A.

William J. Oswald, Division of Sanitary Engineering, University of California, Berkeley, California, 94720, U.S.A.

J. T. Pfeffer, Department of Civil Engineering, University of Illinois, Urbana, Illinois 60801, U.S.A.

Ram Bux Singh, Gobar Gas Research Station, Ajitmal, Etawah, Uttar Pradesh, India.

R. J. Smith, Department of Agricultural Engineering, Iowa State University, Ames, Iowa 50010, U.S.A.

Chapter 7

Waste Treatment and Utilization

All living systems process materials and energy in such a way as to yield a desired end-product or use, plus waste substances. The residues of one system may constitute the raw materials of succeeding systems, although this is not always the case.

Waste may be defined as any material or energy form that cannot be economically used, recovered, or recycled at a given time and place. Under such a definition, wastes could theoretically be disposed of most economically by their discharge to air, water, or soil. However, where human, animal, and plant numbers are large, the direct discharge of untreated liquid, gaseous or solid residues, or wastes frequently leads to severe environmental degradation and even to disease and death in man and other living creatures.

As public recognition of the consequences of environmental pollution has increased, so has the enactment of restrictive antipollution laws. Such laws, together with the increasing cost of raw materials and energy, have led to renewed studies of waste treatment and disposal. These environmental protection laws have also led to increased interest in the development of techniques to recycle and reuse wastes.

Recycling of human, animal, and vegetable wastes has been practiced by man for centuries. These practices have served their purposes, providing, for instance, fertilizer or fuel, but they have often been complicated by the presence of enteric pathogens that have infected the people involved in their handling. Additional pollution problems have arisen more recently because modern industry generates a multitude of nonbiodegradable organic material and heavy metals that find their way into municipal, industrial, and agricultural wastes. Some industrial effluents cause damaging biological effects as they are recycled through the plant and animal food chain. Fortunately, in most recycling processes there has been little adverse effect because recycled toxicants have not entered the food chain.

Some recycled organic materials can be useful as food, feed, crop fertilizer, fermentable substrates, or soil conditioners for nonagricultural land. In recognition of this diversity, it is important to identify the optimal use of recycled organic materials as components of food or feed. Those which are to be recycled for direct refeeding to animals should be handled separately and according to procedures that can be readily controlled.

This chapter focuses primarily on the recycling and utilization of wastes either of biological origin or generated in agricultural processes.

Certain practical but underexploited processes developed for waste and water treatment provide for nutrient and energy reclamation through biological (algal-bacterial) systems. Another largely unexploited but valuable process is the composting of organic wastes. In contrast to burial or incineration, composting enables preservation and reuse of nutrients and minimizes environmental pollution. Recycling of animal wastes through refeeding processed waste to animals, and application of algal-bacterial systems to the treatment and recycling of animal wastes, are viable processes that also appear to be underexploited.

Each of these processes is described in detail in the sections below.

Algal-Bacterial Systems

Algae can both utilize light energy and capture and concentrate nutrients from dilute aqueous solutions. Some algae are capable of growing commensally in an ecosystem with waste-oxidizing bacteria. The results of the commensal metabolism are the release of oxygen and synthesis of bacterial degradation products into new, protein-rich plant material. Algae and bacteria can be used for the treatment and conversion of human and animal wastes into forms useful for fish and animal feeds. It is even possible that algae and bacteria grown on selected vegetable wastes can produce cell protein suitable for human consumption. Algal and bacterial protoplasm are very similar in chemical composition. Both have similar metabolic pathways, although bacteria have more varied metabolisms.

Algal-bacterial processes generally can be divided into two major categories: 1) those designed to oxidize waste, and 2) those designed for optimal production of algae and nutrient recycling.

Liquid Waste Treatment

Two waste treatment processes involving algal-bacterial systems are now available: facultative ponding and integrated ponding, discussed below.

Facultative Ponding In facultative ponding (that is, ponding involving both aerobic and anaerobic treatment), untreated waterborne waste materials are introduced at a bottom center point of a deep (up to 3 m) pond designed to hold the waste for 4–12 weeks, depending on the temperature and concentration of waste material. Shorter holding periods would be possible in the

FIGURE 7.1 Aerial photograph of facultative ponding system in Esparto, California, U.S.A., consisting of ponds in series: two primary, one secondary, one tertiary, and one quaternary. (Photograph courtesy of W.J. Oswald)

torrid zones, with longer holding periods required in temperate zones. Under such conditions, the waste undergoes fermentation. Fermentation products are either given off as gas (such as CH_4 or CO_2), or oxidized by aerobic bacteria that utilize the oxygen produced by algae growing near the surface.

Facultative ponds are usually built in series. Typically, sewage is channeled through four or five successive ponds (see Figure 7.1). Wastes are pumped into the bottom of the first pond, where anaerobic digestion begins. Effluent is removed near the bottom of the first pond and transferred to the bottom of the second pond, where further decomposition (stabilization) occurs through aerobic processes. Cleaner water near the surface of the second pond is in turn transferred to the final two ponds in sequence. The effluent from the last pond should have a low coliform count and be suitable for crop irrigation, except for vegetables to be eaten raw.

Facultative ponding is applicable to most liquid wastes, including domestic and municipal sewage. It is also an appropriate treatment for wastes from vegetable canneries and sugar refineries. In the latter cases, or whenever the loading to a pond in the tropics exceeds 110 kg of Biochemical Oxygen Demand (BOD_5) per ha per day, floating surface mechanical aerators may be necessary. (BOD_5 is the quantity of oxygen required by aerobic microorganisms to oxidize the biologically available organic matter in a waste material during 5 days at 20°C.) At lower temperatures, the loading limit will be correspondingly lower.

The maximum rate of oxygen production by algae in such systems is about 450 kg per ha per day in the tropics. However, because of cloudy days and bacterial turbidity, loadings well under 110 kg per ha per day are recommended for many wastes. Mixing is critical in determining the maximum load.

Major advantages of facultative ponding are low capital costs, low maintenance requirement, good effluent quality, and limited potential for adverse environmental impact. For example, facultative ponds seldom contribute undesirable nitrate or phosphate to the groundwater.

Disadvantages of facultative ponding are: 1) such ponds if overloaded produce foul odors; 2) they are inefficient in nutrient recovery because nitrogen is lost to the atmosphere and most phosphates are precipitated out; and 3) when evaporative loss from the pond exceeds the amount of liquid gained through rainfall, facultative ponds increase in inorganic salt concentration. When this occurs, the salts in the effluent may render it less desirable for the irrigation of certain salt-sensitive crops.

Integrated Ponding In the integrated ponding process (see Figure 7.2), a facultative pond is followed by an algal growth pond. Algal ponds are characterized by high decomposition rates due to high oxygen concentrations produced by the algae. The oxygenated discharge from the algal growth pond is recycled to the surface of the facultative pond. The algal pond is normally about one meter deep and is designed to operate on a holding period of 5–10 days for the waste being treated. The algal pond is equipped with channels and pumps designed to maintain a flow velocity sufficient to bring about the resuspension of algae that have settled to the bottom, where photosynthesis and oxygen production cannot occur (see Figure 7.3).

FIGURE 7.2 Aerial photograph of integrated ponding system at St. Helena, California, U.S.A. The square pond is a primary facultative pond. It is 0.9 ha in area and receives the waste of 3,500 persons. Next to facultative pond (center right) is a 1.82-ha high-rate pond. Lower right is a 0.8-ha algae-settling pond. The two ponds at upper left are disposal ponds. (Photograph courtesy of W. J. Oswald)

FIGURE 7.3 High-rate pond at Modesto, California, U.S.A. The pond is 80 ha and produces an average of 40 t of oxygen per day. Four 100-HP mixing pumps in foreground move water through a 60-m wide × 1.4-m × 3,600-m circulation channel. (Photograph courtesy of W.J. Oswald)

Another desirable aspect of an algal growth pond is the tendency of the pond water to reach a high pH level at about dusk each day as a result of carbon dioxide utilized by the algae. The high pH (~ 9) causes a reduction in bacterial level, and effluents from such high-rate ponds often have low *Escherichia coli* concentrations.

After retention in the algal growth pond, the wastewaters are introduced into the bottom of a third, deep, elongated pond, which serves as a settling pond. Here the algae settle out and a relatively clear algae-free effluent is produced for disposal or discharge to the environment. Integrated pond systems, though more costly than facultative ponds, require less land and produce an effluent superior with respect to both bacterial cell and salt concentrations. Integrated ponds principally consume solar energy; yet they produce an effluent equal in quality to that derived from electrical energy systems in which oxygen is supplied by mechanical aerators.

The disadvantages of integrated ponding systems are their need for solar energy in excess of 200 g cal/cm^2/day and a mild temperature. Occasionally, predators may disrupt the algal population.

Ponding procedures involving algae provide a number of research opportunities, including:

- The application of facultative and integrated ponds to developing-country conditions; and

• The possibility of harvesting feed-grade algae from the secondary or tertiary pond of the series. Where algae harvesting is desirable, it would be feasible to convert an algal pond from an integrated system into an algal production system. This conversion could be accomplished by decreasing the pond depth and recovering the algae by flotation, sedimentation, or straining.

Algal Production

Maximum algal production from domestic sewage and animal wastes is desirable, providing the wastes contain no toxic substances, because it permits conservation of fixed nitrogen in a form useful as animal feed. If there is doubt concerning the quality of the product for feed, it may be usable as a fertilizer or as a fermentation substrate.

After separation from the pond effluent and subsequent drum or spray drying (which can be expensive), or on sand beds (relatively simple and inexpensive), algae constitute a potentially stable product that contains 40-60 percent protein, 10-20 percent carbohydrate, 5-15 percent lipid, 5-10 percent fiber, and 5-10 percent ash. If used daily, moist algae can be dewatered to about 15 percent solids and incorporated with other ingredients such as grain at a concentration of up to 5-10 percent in feeds.

Dry algal protein is up to 80 percent digestible by ruminants. If the material is free of pathogens and toxic substances, it can be used to replace soybean meal, meat, or bone meal in animal, poultry, and fish diets.

Although larger microalgal forms are less common in the environment than unicellular microalgae, they are more desirable for production because they can be harvested by screening and sedimentation. Among the larger microalgal forms, *Spirulina* is the most promising. *Spirulina* cells are large enough to be recovered by simple filtration. In Chad, villagers recover them by using muslin. Dried *Spirulina* resists bacterial degradation and is easily stored. *Spirulina* protein has a satisfactory balance of essential amino acids, with the exception of a slight deficiency in those that contain sulfur. A pilot plant has been set up near Mexico City to collect and process *Spirulina*; about one t per day of dry *Spirulina* is produced and sold as an additive for chicken feed.

Scenedesmus is the most convenient algal genus because it is readily cultured and harvested, particularly when grown under conditions that induce cloning. *Chlorella* species, on the other hand, are less desirable because they are too small to be harvested economically and they are usually eliminated from waste systems through rotifer predation. *Scenedesmus* species are not grazed by rotifers. Infestation by the copepod Cyclops, however, can lead to their eradication within a few days. Cyclops and other *Scenedesmus* grazers may be removed by screening and recirculation of the effluent. *Spirulina* has no comparable predators.

Scenedesmus growth for maximum algal production resembles the algal-ponding process used in waste treatment, except that a lesser pond depth is involved. The waste is introduced into a shallow, channeled growth pond (20–30 cm deep) equipped with paddle wheels to provide a mixing action. Linear flow velocities of 5–15 cm/sec are required. Suitable substrates are:

- Liquid wastes such as domestic sewage effluents;
- Effluents from anaerobic ponds used to treat concentrated plant and animal wastes;
- Digester effluents and residues;
- Effluents from algal and manure fermentation systems used for methane production; and
- Irrigation return flows, urban runoff, and dilute petroleum wastes after the addition of nutrients.

Wastewaters should have a suitable BOD_5, plus an algal growth potential not exceeding 500 mg/1.

An important advantage of algal production systems in conjunction with animal feed lots is that up to 80 percent of the fixed nitrogen and other nutrients are recovered. Yields of up to 60 t/ha/year of dry algal feedstuff may be possible.

Limitations

The algal growth and waste utilization process just described is limited to climatic regions where ambient solar energy is greater than 200 cal/cm^2/day. Another restraint is the requirement that the algal growth potential of the wastewater be sufficiently great to support photosynthetic oxygen production equal to or greater than the BOD_5 of the wastewater. If the oxygen is not produced at a sufficient rate, supplementary oxygen is needed, and the potential yield of algae will be too low to justify the expense of harvesting. The process, however, is readily adaptable to the treatment of residual ("unfeedable") wastes from confined feedlots, since most of the potentially hazardous substances in the wastes can be excluded from such an operation.

In addition to a warm climate and a BOD_5 loading of about 225 kg/ha/day, requirements include level land on which to construct the ponds, a market for a high-protein animal feedstuff (in this case, algae), and sufficient capital to construct the algal growth system.

Algal growth on feedlot wastes poses the risk of possible transmission of disease-causing organisms or toxic substances, unless care is exercised in waste management and selection.

Research Needs

In connection with increasing the use of algal substances, research in the following areas should be emphasized:

- Improved methods of harvesting algae;
- Technology for processing algae to improve digestibility; and
- Possible disease and toxicity hazards.

Composting

Composting is the biological decomposition of organic residues or wastes under controlled conditions to yield a product useful in agriculture.

Although the art of composting is an old one, it has been underexploited. For example, in the maize-growing regions of Mexico, composting is not practiced, despite a great need for organic matter in the soil. In Brazil, São Paulo farmers are very reluctant to use municipal compost supplied to them free of charge, and use it only because of government insistence.

Composting involves the acceleration of microbial decomposition through conditions favorable for microbial reproduction and metabolic activity. Controlling factors are temperature, oxygen supply, moisture, and of course the nature of the substrate.

Temperature There is considerable controversy in temperate climates as to the relative merits of mesophilic (10°- 45°C) versus thermophilic (50°-70°C) composting. In practice, the question is irrelevant, especially in developing countries, since the temperature of a composting mass soon rises to thermophilic levels. This is because of the excess energy generated by bacterial activity combined with the insulating property of the composting mass. High temperatures serve to kill disease-causing organisms as well as fly eggs, larvae, and pupae. Temperature rise is a useful indicator of operational success.

Aeration and moisture content The aerobic approach is followed because higher temperatures are reached thereby and because anaerobic composting produces foul odors. Moisture content and aeration are interdependent. The oxygen used by the microbes comes from air in the spaces between particles of the composting mass. If the spaces are filled with water, air is excluded and aerobic activity is reduced or the process becomes anaerobic. The maximum permissible moisture content varies with the nature of the composted waste. For example, if the bulk of the compost is straw, the maximum permissible moisture content is 80-85 percent. If paper is the major constituent (as in the case of municipal refuse in the United States), 55 percent moisture is the maximum because the paper tends to compact.

Aeration may be accomplished by turning (windrow composting), by mechanical tumbling of the material, or by use of a blower system. Turning, which can be done either manually or mechanically, involves spreading and reforming the windrows. Tumbling can be accomplished by placing the wastes in a rotating drum equipped with vanes or by dropping the material from one level to another.

Substrate The substrate for composting can be almost any organic residue or waste that provides the nutrients required by the microbes. With organic wastes, the proportion of carbon to nitrogen, the major nutrients, may require adjustment. These should be present in a ratio no greater than 30 : 1. At higher ratios, the process is slowed and the quality of the finished product is lowered. At ratios lower than 20 : 1, nitrogen loss can occur through volatilization of ammonia. Examples of nitrogen sources that can be used to adjust the C : N ratio are manures, green plant debris, and animal or fish scraps. Examples of carbon sources are straw, dry vegetable matter, and paper. In most wastes (community and agricultural), phosphorus, potassium, and trace elements are present in sufficient amounts.

Composting is enhanced by uniformity of the particles of substrate. Reducing particle size (grinding) before composting may be advantageous. The optimum particle-size distribution depends upon the materials to be composted. With paper-rich wastes it is in the order of 5 cm. Green vegetables wastes can be larger. In fact, garden debris (excepting woody material) need not be ground.

Any organic waste can serve as a substrate for composting. But care must be taken when human excreta are composted because of the risk that dangerous organisms may not be destroyed.

Organisms The composting process is carried out by a complex mixed population of naturally occurring bacteria. The addition of inocula in composting is normally unnecessary, since the required numbers and variety of microorganisms are already present in the wastes, especially in rural areas.

Advantages

A major advantage of composting is its flexibility with respect to volume of materials handled and degree of mechanization. Composting can range from the individual farm level to a level that can accommodate waste from a village or small town. Sophistication can range from an operation involving manual turning to one in which a complex reactor (digester) is employed.

Another important advantage is that disease-causing organisms are usually rendered harmless during composting. The inactivation may be brought about by high temperature, exhaustion of nutrients, and natural antibiosis.

Perhaps the principal advantage of composting is the production of a product useful in agriculture. Compost can improve the texture of soil, increase its water-holding capacity, and supplement and promote efficient utilization of plant nutrients.

Limitations

Under certain circumstances a potentially important constituent, one which might have a value greater than the compost, might not be reclaimed. For example, it is more economical to recycle paper than to compost it.

A portion of the nitrogen in the wastes is lost in composting. This loss can be reduced by adjusting the carbon : nitrogen ratio of the wastes to a level between 20 : 1 and 30 : 1.

Wastes such as farmyard manure (unless it contains appreciable amounts of straw) must be mixed with a bulking material—a rather difficult task.

The cost and energy involved in turning large quantities of waste during composting may be significant.

A major problem occurs with the composting of untreated human excreta. Extreme care is essential in carrying out the process itself, and certain restrictions must be applied in the use of the product. The product can be safely used on land that is then allowed to lie fallow for at least a year; even the more resistant pathogens are killed during this period.

Another constraint pertains to the carbon : nitrogen ratio of the product. The ratio of carbon : nitrogen must be between 20 : 1 and 30 : 1. At higher carbon levels, the microorganisms growing in the compost preferentially use the nitrogen; this becomes a detriment to plants growing on the land to which the compost is added.

Research Needs

Project operations should be preceded by small-scale "trial-and-error" runs to arrive at useful operational parameters. These trials are needed because of the diversity of waste materials.

Anaerobic Lagoons

Anaerobic lagoons, designed to treat concentrated organic waste, provide a microbial environment in some ways similar to that found in the rumen or intestinal tracts of animals, in sewage sludge digesters, and in the muds and sediments of aquatic areas. Animal wastes are rich in degradable solids and differ considerably from sewage wastes, which are greatly diluted with water. An aerial view of an empty anaerobic lagoon is shown in Figure 7.4.

Properly operating anaerobic lagoons are characterized by an array of microbial associations that ultimately produce methane and carbon dioxide. Three main groups of organisms are involved. The first group degrades and solubilizes fats, proteins, and cellulose. A second group converts these degradation products to a mixture of organic acids and carbon dioxide. The third

FIGURE 7.4 Aerial photograph of anaerobic pond 1.6 ha × 6.2 m deep created to treat the wastes of 2,500 feeder cattle. Steve Marks's feedlot, Zamora, California, U.S.A. (Photograph courtesy of W. J. Oswald)

group utilizes this mix to produce methane. For vigorous fermentation to occur in lagoons, many months may be necessary for maturation. The maturation process may be shortened by the addition of dewatered digested sewage sludge from either a vigorously operating city sewage treatment facility or another functioning lagoon.

The balance among the bacteria may be disturbed by overloading the lagoons with organic material. Imbalance may also occur because of low ambient temperatures. With overloading, increased concentrations of short-chain fatty acids occur, resulting in more substrate than the methanogens can utilize. In the case of low temperatures, the methanogenic population and its rate of metabolism are diminished. As acid concentrations increase, a point is reached where the buffer capacity of the system is overwhelmed and a precipitous drop in pH results. Under acid conditions methanogenesis ceases and the acid- and cold-tolerant group of fermentative organisms continues to make more fatty acids. At high concentrations these acids exert a toxic effect on methane-producing bacteria.

Recovery of an anaerobic lagoon is a sluggish process. It is aided by discontinuing the flow of new waste to the lagoon and bringing the pH to neutrality. A new start may be initiated by adding lime or sodium bicarbonate in amounts calculated from analysis of samples and then waiting for the slowly proliferating methanogens to reestablish themselves in sufficient numbers. The addition of dewatered sludge, with its vigorous population of methanogens, will hasten restoration of the microbial balance.

Anaerobic lagoons respond to warm temperatures with increased rates of catabolism of organic materials and higher populations of microorganisms. Cold weather diminishes rates of organic degradation and reduces microbial numbers. Of the three groups of simultaneously operating microorganisms associated with anaerobic fermentation, the methanogenic bacteria are probably the most sensitive to changes in temperature, and are thus limiting for the fermentation process.

Limitations

Anaerobic lagoons operating at optimum rates of activity require temperatures of 29°-35°C and do best in tropical climates. However, anaerobic fermentation with gas production also occurs in lagoons in temperate climates. Here, lower ambient temperatures are compensated by increasing the pond size by 50 percent in areas of severe winters. Low seasonal temperatures, however, may reduce the numbers of methanogenic bacteria, and their lower rates of metabolism will result in unpleasant odors.

Excavated earthen ponds require relatively nonporous soil to prevent seepage of water. Concern about possible contamination of underground waterways by wastewater has resulted in requirements for lining the basin with bentonite clays and polyphosphates to give an almost impervious seal. Manure ponds in sandy loam soil, however, have been shown to be sealed effectively in less than 6 months by a layer of largely microbial composition.

Research Needs

- More study is needed to characterize the groups of fermentative, hydrogen-producing, acetogenic, and methanogenic microorganisms in anaerobic processes as functions of temperature. If methanogens can be found with higher rates of metabolic activity at lower temperatures, it might be possible to increase the rate of organic waste degradation in cool anaerobic lagoons by adding these bacteria as an augmenting inoculum.
- Additional study will be required for devising an inexpensive method of collecting methane from lagoons to take advantage of a now-wasted energy source. Plastic sheeting, relatively unstable in air and sunlight, might serve as a stable, submerged tent to collect gas from which methane could be separated, or, since algae are not involved and sunlight unnecessary, opaque coverings such as ferrocement could be used.

Recycling Animal Waste by Aerobic Fermentation

Livestock manures are widely utilized as fertilizer and soil conditioners because they contain substantial amounts of the major nutrients needed by agricultural crops. They are also used by some farm families and villages to produce methane for cooking by fermentation (Chapter 6). Another possibility for recycling part of the animal waste is to refeed processed material to the same or other types of animals so that the food value of undigested plant material and microbes is not lost. This has been tested in a microbial processing of livestock waste by lactic fermentation, which produces a silage-like product.

Fresh feedlot cattle-waste solids were separated from the liquid portion. These solids were then combined with each of a number of various cracked grains, mainly maize, in a 1 : 2 ratio and adjusted to 40 percent moisture content. The mixtures were tumbled slowly (0.5 RPM) in a cement mixer at 25°–30°C for 36 hours. The results were a rapid production of acid and control of fetid odor in this aerobic, solid-substrate fermentation, with a final product with an amino acid content 18 percent greater than that of unfermented corn. The organisms in this process came from the waste, not the grain, and conditions favored proliferation of lactic acid bacteria from less than 1 percent of initial total microorganisms to dominant numbers within 12 hours. The acid produced reduced the number of coliform bacteria and other undesirable organisms.

Aerobic culture with substrates of fresh swine waste combined with cracked corn adjusted to 40 percent moisture also resulted in lactic fermentation, with early control of fetid odor and production of a silage-like product in 36 hours. Lactic acid bacteria, indigenous to fresh swine waste, became dominant within 24 hours and produced lactic and other short-chain acids from acetic to valeric. The acidity dropped 2 pH units into the pH 4.2–4.6 range. Although the fermentation product contained 21–39 percent more methionine than maize, when fed to young pigs it was still found inadequate for this amino acid as well as for lysine.

This swine waste fermentation product was fed as the major dietary component to young pigs, hens, and sheep. Pigs showed gain and gain-to-feed ratios diminished by one-third in 13-day trials. Laying hens performed comparably to controls in a 21-day test, and sheep did not discriminate against the fermentation product in a 10-day trial.

Fresh cattle waste aerobically cultured with corn is dominated by lactobacilli. Initially present in small numbers, two-thirds of the lactic acid bacteria are similar to *Lactobacillus fermentum.* After 6 hours, *L. buchneri* dominate and remain high through the 24th hour. In a comparable aerobic swine waste-corn fermentation, more than 98 percent of the *Lactobacillus* sp. initially present were *L. fermentum*, and this organism remained predominant for 144 hours, never dropping below 69 percent of the lactobacilli isolated.

With either swine or cattle waste, yeasts are a major competing group of organisms. If the fermentations are sufficiently aerated, yeasts will increase at the expense of lactic acid bacteria, apparently by inhibiting lactic acid production. The observed change in acid levels may possibly result from utilization of the organic acids by yeasts. The major species of yeast appears to be *Candida krusei.*

Limitations

Aerobic fermentation of cracked cereal grains combined with waste requires tumbling both to mix and to provide oxygen for the microorganisms. However, power requirements are low because the vessels are rotated slowly. Moisture content of fermentation material can be flexible, ranging from 35 to 43 percent. Drier material allows less microbial growth and acid production; excessively wetted material tends to clump. Lysine is the principal limiting amino acid for growing pigs and layer hens in this fermentation product.

In recycling animal waste for its nutrients, the dung of healthy animals is required; diminished disease potential is associated with acid production that kills coliform bacteria. But it is believed that many animal wastes can be refed to livestock without harmful effects to animals or risks to man.

Research Needs

• Decreasing the power requirements for mixing would be helpful. Less power would probably be needed to turn an auger that could mix and aerate this type of fermentation in a stationary, cylindrical vessel with a conical bottom; this impeller design has apparently not been tested in this particular semisolid fermentation.

• The principal limiting amino acid is lysine, and it is desirable to find microorganisms that excrete lysine. However, such organisms may not survive as inoculum in a mixed culture of natural flora.

• Culture techniques are needed to yield more fermentation acid to diminish the disease potential of fermentation products. The effect of aerobic fermentation of waste cereal grain on parasites and viruses is not known, and this represents a potential hazard. Aside from the disease potential, the esthetic and psychological aspects of refeeding processed waste to animals should be studied to assess acceptance of the process by farmers and consumers.

• The consequences of buildup of nonbiodegradable residues as a result of continued recycling should be studied.

• The costs of fresh feed *vs.* costs accrued in collecting and processing wastes need to be determined for each situation. It may be more economical to utilize manures to increase crop yields.

Recycling Animal Waste by Anaerobic Fermentation

Anaerobic culture of fresh cattle waste with ground coastal Bermuda grass or Johnson grass in a ratio of 57 : 43 produces silage by a lactic acid fermentation. Manure and hay are blended and added to the top of an airtight silo; the product removed from the bottom can serve as part of a less costly, adequately nutritious ruminant ration.

The inoculum of lactic bacteria for this fermentation came from feedlot waste and grass. Lactic acid bacteria were isolated from fresh cattle waste in a feedlot and were identified as *Lactobacillus plantarum*, *L. casei* subspecies *casei*, *L. casei* subspecies *alactosus*, and *L. fermentum.* Uncut grass has few lactic acid organisms, but harvesting is an important mechanism for spreading these microbes, which are usually associated with decayed material in contact with the soil, providing numbers comparable with those in feedlot waste.

Limitations

In work that has continued since 1962, the potential disease hazards of refeeding of cattle waste processed by bacterial fermentation appear limited, as judged by the absence of reports of infection.

Diminished risks of disease are associated with maintenance of apparently healthy herds; for example, isolation pens are used with new feeder stock. Reduced risks are also linked with the process of ensiling, which involves lactic acid bacteria, and produces largely lactic and acetic acids that increase acidity to near pH 4.0. The effect of these acids on enteric pathogens was demonstrated by inoculation with each of 27 serotypes of *Salmonella* into separate laboratory silos. No *Salmonellae* survived in the manure-blended ration, whereas 25 or 27 *Salmonella* serotypes were recovered in silos that contained only manure. In another study, eggs of nematodes in manure combined with coastal Bermuda grass hay and ensiled for 4 weeks demonstrated that parasitic larvae were absent in the finished product.

Research Needs

Cattle being fed or treated with antibiotics or related substances may produce waste containing undegraded and diluted drugs. It is known that low concentrations of antibiotics taken by an animal may favor the development of microorganisms resistant to the antibiotic in use. It is not known, however, what effect this fermentation process may have on inactivating antibiotics or other therapeutic chemicals used with animals, and research is needed to determine this.

References and Suggested Reading

Algal-Bacterial Systems

Gloyna, E. F.; Malina, J. F.; and Davis, B. M., eds. 1976. *Water resources symposium.* Vol. 2: *Ponds as a wastewater treatment alternative*. Austin: University of Texas, Center for Research in Water Resources.

Laskin, A. I., and Lechevalier, H., eds. 1978. *CRC handbook of microbiology*. 2nd edition, Vol. II: *Fungi, algae, protozoa and viruses.* West Palm Beach, Florida: CRC Press.

Oswald, W. J.; Lee, E. W.; Adan, B.; and Yao, K. H. 1978. New wastewater treatment method yields a harvest of saleable algae. *WHO Chronicle* 32:348-350.

Composting

Compost Science/Land Utilization. Emmaus, Pennsylvania: Rodale Press, Inc.

Golueke, C. G. 1972. *Composting.* Emmaus, Pennsylvania: Rodale Press, Inc.

_____. 1977. *Biological reclamation of solid wastes.* Emmaus, Pennsylvania: Rodale Press, Inc.

_____, and McGauhey, W. J. 1952. *Reclamation of municipal refuse by composting.* Sanitation Engineering Research Laboratory Technical Bulletin, No. 9. Berkeley, California: University of California.

Anaerobic Lagoons

Bryant, M. P. 1979. Microbial methane production—theoretical aspects. *Journal of Animal Science* 48 (1): 193-201.

Kirsch, E. J., and Sykes, R. M. 1971. Anaerobic digestion in biological waste treatment. *Progress in Industrial Microbiology* (London) 9:155-237.

Miner, J. R., and Smith, R. J., eds. 1975. *Livestock waste management with pollution control.* Midwest Plan Service Series, No. MWPS-19. Ames, Iowa: Iowa State University.

Zeikus, J. G. 1977. The biology of methanogenic bacteria. *Bacteriological Reviews* 41:514-541.

Recycling Animal Wastes by Aerobic Fermentation

Fontenot, J. P., and Webb, K. E., Jr. 1975. Health aspects of recycling animal wastes by feeding. *Journal of Animal Science* 40:1267-1277.

Rhodes, R. A., and Orton, W. L. 1975. Solid substrate fermentation of feedlot waste combined with feedgrain. *Transactions of the American Society of Agricultural Engineers (ASAE)* 18:728-733.

Smith, L. W.; Calvert, C. C.; Frobish, L. T.; Dinius, D. A.; and Miller, R. W. 1971. *Animal waste reuse—nutritive value and potential problems from feed additives*. ARS-44-224. Washington, D.C.: U.S. Department of Agriculture.

Weiner, B. A. 1977. Fermentation of swine waste-corn mixtures for animal feed: pilot-plant studies. *European Journal of Applied Microbiology* 4:59-65.

Recycling Animal Wastes by Anaerobic Fermentation

Anthony, W. B. 1971. Cattle manure as feed for cattle. In *Livestock waste management and pollution abatement: Proceedings of the International Symposium on Livestock Waste,* April 19-22, 1971, Ohio State University, Columbus, Ohio, pp. 293-296. St. Joseph, Michigan: American Society of Agricultural Engineers.

McCaskey, T. A., and Anthony, W. B. 1975. Health aspects of feeding animal waste conserved in silage. In *Managing livestock wastes: Proceedings of the Third International Symposium on Livestock Wastes*, April 21-24, 1975, University of Illinois, Urbana-Champaign, Illinois, pp. 230-233. ASAE Publication 275. St. Joseph, Michigan: American Society of Agricultural Engineers.

Research Contacts

Algal-Bacterial Systems

J. Benemann, University of California, Richmond Field Station, 1301 South 46th Street, Richmond, California 94804, U.S.A.

E. F. Gloyna, University of Texas, Austin, Texas 78712, U.S.A.

W. J. Oswald, Division of Sanitary Engineering, University of California, Berkeley, California 94720, U.S.A.

Composting

Jerome Goldstein, Editor, *Compost Science/Land Utilization*. Box 351, Emmaus, Pennsylvania 18049, U.S.A.

C. G. Golueke, Cal Recovery Systems, Inc., 160 Broadway, Suite 200, Richmond, California 94804, U.S.A.

Anaerobic Lagoons

M. P. Bryant, Department of Dairy Science and Microbiology, University of Illinois, Urbana, Illinois 61801, U.S.A.

Raymond C. Loehr, College of Agriculture and Life Sciences, Cornell University, Ithaca, New York 14853, U.S.A.

J. Ronald Miner, Agricultural Engineering Department, Oregon State University, Corvallis, Oregon 97331, U.S.A.

William J. Oswald, Division of Sanitary Engineering, University of California, Berkeley, California 94720, U.S.A.

B. A. Weiner, Fermentation Laboratory, Northern Regional Research Center, U.S. Department of Agriculture, Agricultural Research Service, 1815 N. University, Peoria, Illinois 61604, U.S.A.

Recycling Animal Wastes by Aerobic Fermentation

J. P. Fontenot, Department of Animal Science, Virginia Polytechnic Institute and State University, Blacksburg, Virginia 24061, U.S.A.

L. W. Smith, Biological Waste Management Laboratory, Agricultural Environmental Quality Institute, Science and Education Administration, U.S. Department of Agriculture, Beltsville, Maryland 20705, U.S.A.

B. A. Weiner, Fermentation Laboratory, Northern Regional Research Center, U.S. Department of Agriculture, Agricultural Research Service, 1815 N. University, Peoria, Illinois 61604, U.S.A.

Recycling Animal Wastes by Anaerobic Fermentation

W. Brady Anthony, Animal and Dairy Sciences Department, Alabama Agricultural Experiment Station, Auburn University, Auburn, Alabama 36830, U.S.A.

J. P. Fontenot, Department of Animal Science, Virginia Polytechnic Institute and State University, Blacksburg, Virginia 24061, U.S.A.

L. W. Smith, Biological Waste Management Laboratory, Agricultural Environmental Quality Institute, U.S. Department of Agriculture, Science and Education Administration, Beltsville, Maryland 20705, U.S.A.

B. A. Weiner, Fermentation Laboratory, Northern Regional Research Center, U.S. Department of Agriculture, Agricultural Research Service, 1815 N. University, Peoria, Illinois 61604, U.S.A.

Chapter 8

Cellulose Conversion

Cellulose is the earth's most abundant renewable raw material, with about 10–15 t per person produced annually by plants. Most of this cellulose occurs in intimate association with a complex plant structural material called lignin. The resulting lignocelluloses are by far the most prevalent renewable organic materials available for microbial—or other—conversions.

Cellulose that is not lignified (for instance, nonwoody aquatic plant tissues, some papers, residues from chemical pulp mills, and certain natural fibers such as cotton) may be available in sufficient quantity in some locales to be considered for microbial conversions. A wide range of microorganisms can degrade cellulose. Far fewer species can degrade the natural lignocelluloses because lignin limits access to the cellulose. In fact, at present, the only currently operative means for converting unmodified lignocellulosics biologically is through the production of various mushrooms, which are a good source of protein for human consumption.

Pretreatment to disrupt or destroy the lignin barrier permits use of a broader range of microorganisms. Both physical and chemical pretreatments have been developed, but the former requires large amounts of energy. Chemical pretreatment might be attractive in some situations, however, and two that are now under investigation seem promising. In one, lignocellulose complex is treated with concentrated phosphoric acid, in the other the lignocellulose residues are treated with sulfur dioxide gas. In both cases this is followed by neutralization with a base.

For some organisms it is necessary to disrupt the lignocellulose complex and remove the lignin; for others removal of lignin is not necessary. Pretreatment using easily manipulated reagents and unsophisticated equipment is necessary for a cottage industry. Ambient temperature processing with mineral acids or alkalis followed by neutralization would meet these requirements but adds to the cost. Certain fungi, such as the mushrooms, decompose lignin and some can be used effectively for pretreating lignocelluloses.

Table 8.1 lists nine microorganisms or processes that are either promising or already in commercial use for the conversion of cellulose or lignocellulose. As shown in Table 8.2, five of these require cellulose or pretreated lignocellulose for efficient conversion. The processes vary not only in the type of substrate and the product, but also in the degree of sophistication and in the present state of knowledge and development. For each process, Table 8.2 lists a specific organism(s) along with some of the conditions related to its use. In the following pages, each process is discussed in more detail.

Volvariella Species

"Padi-straw" mushrooms (*V. volvacea*, shown in Figures 8.1 and 8.2, and *V. esculenta* and *V. displasia*) are cultivated on rice straw and similar materials in the Orient and Africa. They are of increasing commercial importance, but are also traditionally cultivated by individuals. They show promise of greatly expanded use in grain-growing regions of the tropical world. Production involves simply inoculating water-soaked straw in flat beds, maintaining moisture at optimum levels, and harvesting the several crops of mushrooms. The mushrooms may be dried for storage and later use. The spent straw is used to inoculate fresh beds and is probably also used as animal feed.

Before 1970, rice straws were practically the only material used for the preparation of the medium for the mushroom. Recently, a number of other materials such as water hyacinth, oil-palm nut pericarp, cotton, and banana leaves have been shown to be satisfactory culture material. Undoubtedly many other lignocellulosic agricultural residues could also be used satisfactorily.

Volvariella is primarily a fungus of the tropics and subtropics, the areas that include most of the developing countries. These are also the areas in which land is often considered to be the limiting factor in the production of food. In the case of mushrooms cultivated on agricultural residues, land ceases to be an important factor. According to recent data, 1 m^2 of growth space can produce 586 kg of mushrooms per year based on two crops per month.

Limitations

There are no important limitations to the cultivation of *Volvariella* species within the environmental range of growth.

Research Needs

To support increased growth of *Volvariella* species for food, the following research efforts are needed:

- Determination of the best species and strains for given locations and substrates;
- Determination of the optimum environment for each species; and
- Evaluation of various substrates for maximum yields of the mushrooms.

TABLE 8.1 Products of Cellulose- or Lignocellulose-Utilizing Microorganisms

Microorganism	Product	Present Status
Volvariella volvacea	Human food (mushrooms); animal feed	Some commercial use
Lentinus edodes	Human food (mushrooms); animal feed	Used commercially
Pleurotus sp.	Human food (mushrooms); animal feed	Used commercially
Thermoactinomycetes sp. and other thermophilic actinomycetes	Human food (SCP); animal feed	Under research
Phanerochaete chrysosporium	Delignified cellulose for use as feed, fiber, or further conversions	Under research
Trichoderma reesei	Cellulases for converting cellulose to sugars; animal feed (SCP)	Under development
Clostridium thermocellum	Cellulases for converting cellulose to sugar; ethanol, acetate, lactate, and H_2; animal feed (SCP)	Under research
Pseudomonas fluorescens var. *cellulosae* and similar bacteria	Animal feed; cellulases for converting cellulose to sugars	Under research
Thermophilic *Sporocytophaga*	Animal feed; ethanol, acetic acid	Under research

Lentinus edodes

The "shiitake" mushroom has been cultivated and used as human food for centuries in China and Japan, where it is commercially produced in what now is a multimillion dollar industry. It is not used much in most developing countries, nor is it popular in the West where the common champignon, *Agaricus bisporus* (*A. brunnescens*), is the mushroom of commerce. *L. edodes* (Figure 8.3) has an important advantage over *A. bisporus* in that it can be cultivated on wood, mainly but not exclusively on oak (Figure 8.4). Thus, it has potential for the direct bioconversion of lignified residues and low-quality wood into fungal protein. Wood decayed by *L. edodes* is quite digestible by ruminants, although this potential use of the organism has received little attention.

In general it takes 1½–3 years for the production of the fruit bodies after inoculation in log wood. A new method, involving direct injection of the liquid spawn into log wood, has shortened the fruiting time to about 6 months. With this method, the number of spawning points can be increased.

TABLE 8.2 Process Characteristics for Microorganisms Utilizing Lignocellulose or Cellulose

Microorganism	Sterility Required	Substrate	Control of Environment	Process Temperature	Time Required
Volvariella sp.	No	Lignocellulose (straw, etc.)	No*	Ambient (high temperature and humidity)	Weeks
Lentinus edodes	No	Lignocellulose (wood)	No*	Ambient	Months
Pleurotus sp.	No	Lignocellulose (straw, wood, etc.)	No*	Ambient	Weeks
Trichoderma reesei	Yes	Cellulose or pretreated lignocellulose	Yes	Ambient	Days
Thermoactinomycetes sp. and other thermophilic actinomycetes	No	Cellulose or pretreated lignocellulose	Yes	55°-65°C	Days
Phanerochaete chrysosporium	No (?)	Lignocellulose (wood)	Yes	Ambient to 40°C	Weeks
Clostridium thermocellum	Yes (?)	Cellulose or pretreated lignocellulose	Yes	55°-65°C	Days
Pseudomonas fluorescens var. *cellulosae*	Yes	Cellulose; pretreated lignocellulose; lignocellulose (?)	Yes	Ambient	Days
Thermophilic *Sporocytophaga*	No	Cellulose or pretreated lignocellulose	Yes	55°-65°C	Days

*Control is necessary, however, for efficient long-term production.

FIGURE 8.1 Fruiting bodies of *Volvariella volvacae* on straw-cotton waste compost. (Photograph courtesy of S. T. Chang)

FIGURE 8.2 The "button" (left) and "egg" stages of *Volvariella volvacea.* (Photograph courtesy of S. T. Chang)

FIGURE 8.3 *Lentinus edodes* fruiting one year after inoculation on bolts of oak wood in Japan. (Photograph courtesy of Y. Hashioka)

FIGURE 8.4 Typical arrangements of bolts of oak wood for cultivation of *Lentinus edodes* (Japan). (Photograph courtesy of Y. Hashioka)

It also reduces the amount of manual labor and minimizes the loss of wood.

The shiitake mushroom is quite perishable, and the best way to preserve its volatile compounds, amino acids, vitamin B content, and texture is by freeze drying. Freeze-dried mushrooms are very close in composition to the original fresh samples. Most of these mushrooms, however, are simply air-dried for storage and marketing.

Limitations

L. edodes may not be suitable or efficient for use with many available woods. No other limitations are expected within its environmental range.

Research Needs

Further study should be focused on:

- Evaluation of available wood species;
- Selection of best strains for specific substrates; and
- Evaluation of *L. edodes* wood residue as a ruminant feed.

Pleurotus Species

The "oyster" mushrooms (*Pleurotus ostreatus* and *P. sajor-caju*) and other species (*P. florida, P. eryngii, P. cornucopiae,* and *P. cystidiosus;* Figure 8.5), like *L. edodes*, preferentially decompose lignin, although they also utilize cellulose and other carbohydrate polymers in wood. As many as three successive

FIGURE 8.5 Fruiting bodies of *Pleurotus cystidiosus,* a commercial and popular mushroom in Taiwan. (Photograph courtesy of J. T. Peng)

harvests on a single substrate batch have been reported. These species have the potential for converting sawmill residue and other low-value wood into protein-rich food for human consumption. All are currently used as food. *P. cornucopiae* is grown commercially in Japan and *P. ostreatus* is grown commercially in Eastern and Western Europe, but they apparently are little used in developing countries. *P. ostreatus* and *P. florida* have temperature optima near 30°C, making them promising for tropical applications. All can be cultivated on straw and on mixtures of sawdust, grain, manure, food-processing wastes, and similar substrates with an added nitrogen source such as commercial fertilizer. A large spawn inoculum eliminates the need for sterilization as the mycelia quickly take over. Yields vary; the highest reported is 100 percent (1 kg per kg of dry substrate) on banana pseudostems.

Limitations

Substrates usually need to be pasteurized unless a large inoculum is used. There are no other important limitations within a suitable environmental range.

Research Needs

The study of *Pleurotus* sp. should be concentrated on:

- Further improvement of cultivation conditions; and
- Selection of the best strains for each location and substrate.

Thermoactinomyces Species

The thermophilic cellulolytic and starch-utilizing actinomycetes provide a unique opportunity for development of a cottage industry for the production of single-cell protein for food or animal feed. They can be grown on high solids (40–60 percent moisture) under conditions of aeration and temperature in which most contaminants cannot compete.

At 40–60 percent moisture, the biomass is a thick paste or a damp, friable solid. It can be spread as a thin layer in trays or on fine mesh screens in an incubator. The screens are preferable to trays because both sides are exposed to air. It is necessary to control temperature, moisture content, pH, and oxygen content during growth. The organisms tolerate variations in temperature (50°–65°C), moisture (40–60 percent), pH (6.8–8.5), and oxygen (1–20 percent).

All plants and plant wastes are potential substrates for these organisms.

Starchy materials such as cassava, banana, potato, and corn starch can be used directly, while cellulosic materials require pretreatment. The extent and kind of pretreatment depends upon the substrate. Waste newspaper usually contains lignocellulose and may require pretreatment. Soft woods contain more lignin and may require more extensive pretreatment than hardwoods. The lignin content of all fibrous plants increases with age. Therefore, young succulent plants may be utilized without pretreatment, while mature plants of the same species will require pretreatment. Removal of the lignin is not necessary for cellulose utilization and growth, but the lignocellulose complex must be ruptured, so cellulolytic enzymes can penetrate and hydrolyze the cellulose. Without pretreatment, cellulose utilization is much less complete.

Using this organism on various starch and cellulose substrates would be labor intensive but would require minimal capital and technology. The process is no more complicated than making cheese or wine. It can be carried out under clean but nonsterile conditions. However, an environment should be selected in which undesirable contaminants cannot grow.

Thermophilic actinomycetes such as *Thermoactinomyces* sp. have the following characteristics, making them good candidates for a cottage industry:

- They grow at 55°–65° C. Twenty-four hours or more of growth at this temperature range results in a pasteurized product in which most known pathogens would be destroyed.
- They grow rapidly, with a minimum of growth requirements. Fermentation should be complete within a few days.
- If proper control of temperature and moisture is exercised, thermophilic actinomycetes will be the predominant if not the only organism present. Only actinomycetes, thermophilic bacteria, and a few algae grow above 55°C. If the moisture content is kept in the range of 40–60 percent, other bacteria cannot compete effectively. Thermophilic algae will not be a problem in dark growth chambers.
- Nutrients and seed culture can be added from a prepackaged mix, just as yeast is now prepared for bread and wine making.
- A variety of inexpensive organic and inorganic nitrogen sources can be used.
- Temperature and moisture can be manually controlled by moving trays to different incubators or to different parts of the same incubator.

Limitations

Thermoactinomyces sp. does not efficiently utilize lignin-cellulose complex found in most plants. A pretreatment of the substrate may be needed in some cases.

"Farmer's lung" is a possible occurrence; it is caused by inhalation of large numbers of spores, and is an allergic response rather than an infection caused by colonization of the respiratory tract. The symptoms are categorized as hyperactivity pneumonitis or extrinsic allergic alveolitis. If conditions are maintained so that the product remains moist, spore inhalation can be minimized. Thermophilic actinomycetes have not been reported to produce aflatoxins or other mycotoxins, as have many of the higher fungi.

Research Needs

The following research efforts should be emphasized:

- Composition analysis of the single-cell protein product;
- Nutritional evaluation of the product;
- A detailed description of the construction and operation of a low-technology incubator;
- Preparation of a culture and nutrient packet; and
- A description of methods for use and preservation of the product.

Phanerochaete chrysosporium

A ubiquitous wood-decay fungus that inhabits the northern hemisphere is called, variously, *Peniophora "G," Chrysosporium pruinosum, C. lignorum, Sporotrichum pulverulentum*, *S. pruinosum*, and *P. chrysosporium.* It is one of the organisms most damaging to stored wood chips, and, like *Pleurotus* sp. and *L. edodes*, decomposes all components of wood. Among the several hundred species of lignocellulose-destroying fungi, *P. chrysosporium* is unusual in that: 1) it produces copious quantities of asexual spores, making it easy to handle; 2) it is thermotolerant, growing optimally at 35°–40°C, but also growing well at 25°C; 3) it grows very rapidly and is an aggressive competitor; and 4) it decomposes lignin as rapidly as any organism thus far studied. Therefore, *P. chrysosporium* has been selected for detailed study of lignin degradation and the microbial processing of wood.

This fungus should be studied further for converting wood-processing residues and other lignified wastes. The selective degradation of lignin caused by the fungus increases rumen digestibility, and the fungal mycelium adds protein. Partial decay of wood wastes by *P. chrysosporium* should render them suitable for ruminant feed or for further digestion to sugars by cellulolytic enzymes (see discussion of *Trichoderma reesei* below) or by bacteria. The fungus has been fed to fish and rats as the sole protein source with no adverse effects.

Limitation

Some substrates might have to be pasteurized for successful growth of the organism.

Research Needs

Further study on *P. chrysosporium* should have the following goals:

- Determining optimum conditions for delignification of specific substrates;
- Evaluating properties of products;
- Selecting superior strains and achieving genetic improvement; and
- Establishing sterility requirements.

Trichoderma reesei

A number of fungi are cellulolytic, but only a few produce cell-free enzymes in sufficient quantity to be of value in degrading cellulose. *Trichoderma reesei* forms a stable cellulase system that is capable of extensive degradation of cellulose.

T. reesei grows rapidly on simple media and does not require supplemental growth factors. In agitated culture, it produces short mycelial threads; rarely does it form pellets on carbohydrates. It is a strong acid producer and will grow under pH conditions as low as 2.5; during actual enzyme production, the medium can be adjusted to pH 3, thus minimizing contamination. Media can be inoculated with a spore suspension or with a small volume of cellulose-containing mycelia.

To obtain the highest yield of enzyme, interfering substances such as lignin must be removed and the cellulose pretreated.

The cellulases of *T. reesei* have been studied more extensively than those from other organisms. There is extensive literature on conditions of growth for enzyme production, enzyme isolation and purification, and properties of isolated enzymes. If a commercial process for large-scale use of cellulase enzymes is developed, it will probably use *T. reesei.*

Little research has been done on SCP production by *T. reesei.* The protein content and amino acid profile are similar to the microfungi in that the protein is limited in sulfur-containing amino acids. The value of the protein from *T. reesei* in human nutrition has not been reported.

Limitations

The use of *Trichoderma reesei* will most likely be limited to the production of enzymes from pretreated cellulosic material. This will require fermenters and fermenter technology that may not be available in some developing countries. Native lignocellulose would require pretreatment, such as delignification.

If *T. reesei* is grown under nonsterile conditions there is a danger of contamination by mycotoxin-producing fungi.

Research Needs

Many basic and pilot-plant studies have been completed. But more work will be needed, especially on:

- Production of cellulase from *Trichoderma reesei* by the koji process; and
- Methods for increasing enzyme production, substrate-enzyme susceptibility, and enzyme recovery after use.

Other Species*

Some gram-negative, aerobic bacteria such as those in the genera *Pseudomonas* and *Xanthomonas* as well as gram-positive *Cellulomonas* species utilize cellulose but do not utilize lignin; therefore, some form of pretreatment before fermentation is required. These bacteria grow rapidly at room temperature (20°-30°C), are obligate aerobes, and usually require yeast extract or some growth factors. These nonspore-forming bacteria are readily digested by livestock. The amino acid composition is similar to that of other bacteria and constitutes a source of nutritional protein.

Co-fermentation of mesquite wood with *P. fluorescens* and a yeast, *Candida utilis*, has been conducted. The protein yield of *Cellulomonas* has been increased by co-fermentation of cellulose with *Alcaligenes faecalis* and with *Cellulomonas flavigena* and *Xanthomonas campestris*. Apparently, co-fermentation with a noncellulolytic organism increases the rate of utilization of soluble sugars produced by hydrolysis of cellulose.

Limitations

The bacteria mentioned above grow at pH levels of 6.5-8.0 and the fermentations are subject to contamination by noncellulolytic bacteria as well as pathogens; therefore, aseptic conditions must be maintained throughout the fermentation. The need for pretreatment, either chemical or biological, in-

*See also discussion of SCP production in Chapter 2.

creases the cost of protein produced. These bacteria are small (about $1\mu \times 0.5\mu$) and must be harvested by differential centrifugation or recovered on very fine filters.

Research Needs

Research is needed to:

- Determine the effect of lignin on cellulose utilization; and
- Evaluate pretreatment needs for specific substrates such as bagasse, waste paper, and various woods.

References and Suggested Reading

Volvariella Species

Chang, S. T. 1965. How to grow straw mushrooms. *Quarterly Journal of the Taiwan Museum (Taipei)*. 18:477-487.

——. 1977. *The straw mushroom as a good source of food protein in Southeast Asia.* Paper presented at the Fifth International Conference on Global Impacts of Applied Microbiology, November 21-25, 1977, Bangkok, Thailand.

——, and Hayes, W. A. 1978. *The biology and cultivation of edible mushrooms.* New York: Academic Press.

Chua, S. E., and Ho, S. Y. 1973. Cultivation of straw mushrooms. *World Crops* 25:90-91.

Gray, W. D. 1970. *The use of fungi as food and in food processing.* West Palm Beach, Florida: CRC Press.

Ho, Ming-shu. 1972. Straw mushroom cultivation in plastic houses. *Mushroom Science* 8:257-263.

Singer, R. 1961. *Mushrooms and truffles.* Bedfordshire, England: Leonard Hill Books, distributed in the United States by John Wiley and Sons (World Crop Books), New York.

Lentinus edodes Species

Akiyama, H.; Akiyama, R.; Akiyama, I.; Kato, A.; and Nakazawa, K. 1974. The new cultivation of shiitake in a short period. *Mushroom Science* 9:423-434.

Gray, W. D. 1970. *The use of fungi as food and in food processing.* West Palm Beach, Florida: CRC Press.

Singer, R. 1961. *Mushrooms and truffles.* Bedfordshire, England: Leonard Hill Books, distributed in the United States by John Wiley and Sons (World Crop Books), New York.

Pleurotus Species

Block, S. S., *et al.* 1959. Experiments in the cultivation of *Pleurotus ostreatus. Mushroom Science* 4:309-325.

Gray, W. D. 1970. *The use of fungi as food and in food processing.* West Palm Beach, Florida: CRC Press.

Kaneshiro, T. 1976. Lignocellulosic agricultural wastes degraded by *Pleurotus ostreatus. Developments in Industrial Microbiology* 18:591-597.

Singer, R. 1961. *Mushrooms and truffles.* Bedfordshire, England: Leonard Hill Books, distributed in the United States by John Wiley and Sons (World Crop Books), New York.

Zadrazil, F. 1976. The ecology and industrial production of *Pleurotus ostreatus, P. florida, P. cornucopiae*, and *P. eryngii. Mushroom Science* (London) 9 (Part-1): 621-652.

Thermoactinomyces Species

Bellamy, W. D. 1974. Single cell proteins from cellulosic wastes. *Biotechnology and Bioengineering* 16:869.

——. 1976. Production of single-cell protein for animal feed from lignocellulose wastes. *World Animal Review* 18:39.

——. 1977. Cellulose and lignocellulose digestion by thermophilic actinomyces for single-cell protein production. *Developments in Industrial Microbiology* 8:249-254.

Blyth, E. 1973. Farmer's lung disease in actinomycetales. In *Actinomycetales: characteristics and practical importance*, G. S. Sykes and F. A. Skinner, eds., pp. 261-276. New York: Academic Press.

Crawford, D. L. 1974. Growth of *Thermomonospora fusca* on lignocellulose pulps of varying lignin content. *Canadian Journal of Microbiology* 20:1069-1072.

——; E. McCoy; J. M. Harkin; and P. Jones. 1973. Production of microbial protein from waste cellulose by *Thermomonospora fusca*, a thermophilic actinomycete. *Biotechnology and Bioengineering* 14:833-843.

Gray, W. D. 1970. *The use of fungi as food and in food processing.* West Palm Beach, Florida: CRC Press.

Hesseltine, C. W. 1972. Solid state fermentations. *Biotechnology and Bioengineering* 14:517-532.

Imrie, F. 1975. Single-cell protein from agricultural wastes. *New Scientist* 66:458.

Stutzenberger, F. J. 1972. Cellulolytic activity of *Thermomonospora curvata*: nutritional requirements for cellulase production. *Applied Microbiology* 24:77-82.

Terui, G.; Shibasaki, I.; and Mochiguki, T. 1958. Studies on high-heap aeration process as applied to some industrial fermentations: II. General description of the improved process. *Osaka University Technology Reports* 3:214.

Phanerochaete chrysosporium

Ander, P., and Eriksson, K.-E. 1977. Lignin degradation and utilization by microorganisms. *Archives of Microbiology* 109:1-15.

Burdsall, H. H., Jr., and Eslyn, W. E. 1974. A new *Phanerochaete* with a *Chrysosporium* imperfect state. *Mycotaxon* 1:123-133.

Eriksson, K.-E., and Pettersson, B. 1972. Extracellular enzyme system utilized by the fungus *Chrysosporium lignorum* for the breakdown of cellulose. In *Biodeterioration of materials: Proceedings of the International Biodeterioration Symposium, 2nd, Lunteren, The Netherlands.* A. Harry Walters and E. H. Hueck-Van Der Plas, eds., Vol. 2, pp. 116-120. New York: John Wiley and Sons.

Hofsten, B. V., and Hofsten, A. V. 1974. Ultrastructure of a thermotolerant basidiomycete possibly suitable for production of food protein. *Applied Microbiology* 27:1142-1148.

Kirk, T. K.; Yang, H. H.; and Keyser, P. 1978. The chemistry and physiology of the fungal degradation of lignin. In *Developments in Industrial Microbiology, Proceedings of the Annual Meeting, August 21-26, 1977, Michigan State University, Lansing, Michigan*, L. A. Underkofler, ed., pp. 51-61. Arlington, Virginia: American Institute of Biological Sciences.

Trichoderma reesei

Gaden, E. L., Jr.; Mandels, M.; Reese, E. T.; and Spano, L. A., eds. 1976. *Enzymatic conversion of cellulosic materials: technology and applications.* New York: John Wiley and Sons.

Mandels, M., and Weber, J. 1969. The production of cellulases. In *Cellulases and their application.* Advances in Chemistry Series, No. 95, pp. 391-414. Washington, D.C.: American Chemical Society.

Other Species

Bruil, C., and Kushner, J. J. 1976. Cellulase induction and the use of cellulose as a preferred growth substrate by *Cellvibrio gilvus. Canadian Journal of Microbiology* 22:1777-1787.

Dunlop, C. E. 1975. Production of single-cell protein from insoluble agricultural wastes by mesophiles. In *Single-cell protein II*, S. R. Tannenbaum and D. E. Wang, eds., pp. 244-267. Cambridge, Massachusetts: Massachusetts Institute of Technology Press.

Han, Y. W., and Callihan, C. D. 1974. Cellulose fermentation: effect of substrate pretreatment on microbial growth. *Applied Microbiology* 27:159-165.

Thayer, D. W. 1976. A submerged culture process for production of cattle feed from mesquite wood. *Developments in Industrial Microbiology* 17:1779-1789.

Research Contacts and Culture Sources

Volvariella Species

Romeo V. Alicbusan, Science Research Supervisor and Head, Microbiological Research Department, National Institute of Science and Technology, Manila, The Philippines.

S. T. Chang, Department of Biology, The Chinese University of Hong Kong, Shatin, New Territories, Hong Kong.

Yoshio Hashioka, 2337 Shinkano Naka, Kagamigahara City, Gifu, Japan

James P. San Antonio, Genetics and Germ Plasm Institute, Vegetable Laboratory, BARC-W, U.S. Department of Agriculture, Science and Education Administration, Beltsville, Maryland 20705, U.S.A.

Lung-chi Wu, Campbell Institute for Agricultural Research, Napoleon, Ohio 43545, U.S.A.

Pleurotus Species

Gerlind Eger, Institut fur Pharmazeutische Technologie, Universitat Marburg, Marbacher Weg 6, 355 Marburg, West Germany.

S. C. Jong, Mycology Department, American Type Culture Collection, 12301 Parklawn Drive, Rockville, Maryland 28052, U.S.A.

J. T. Peng, Mushroom Research Laboratory, Taiwan Agricultural Research Institute, Taipei, Taiwan, R.O.C.

James P. San Antonio, Genetics and Germ Plasm Institute, Vegetable Laboratory, BARC-W, U.S. Department of Agriculture, Science and Education Administration, Beltsville, Maryland 20705, U.S.A.

Thermoactinomyces Species

Cellulose- or starch-utilizing thermophilic actinomycetes can easily be isolated from any compost pile by culturing on starch agar or cellulose agar plates at 55°- 65°C.

W. D. Bellamy, Department of Food Science, Cornell University, Ithaca, New York, 14853, U.S.A.

D. L. Crawford, Department of Bacteriology and Biochemistry, University of Idaho, Moscow, Idaho 83843, U.S.A.

Phanerochaete chrysosporium

Karl-Erik Eriksson, Swedish Forest Products Research Laboratory, Noc 5064, S-114 86 Stockholm, Sweden.

T. Kent Kirk, Forest Service, U.S. Department of Agriculture, Forest Products Laboratory, P.O. Box 5130, Madison, Wisconsin 53705, U.S.A.

Trichoderma reesei

Elwyn Reese, Food Science Laboratory, U.S. Army, Natick Research and Development Command, Natick, Massachusetts 01760, U.S.A.

Other Species

V. R. Srinivasan, Department of Microbiology, Louisiana State University, Baton Rouge, Louisiana 70803, U.S.A.

D. W. Thayer, Department of Biological Sciences and Food Nutrition, Texas Technological University, Lubbock, Texas 79409, U.S.A.

Chapter 9

Antibiotics and Vaccines

The production of antibiotics and vaccines is a further type of microbial processing that may be of great importance to developing countries.

Antibiotics are antimicrobial substances produced by living microorganisms; many are used therapeutically and at times prophylactically in the control of infectious diseases. They act by inhibiting the growth of the infecting organisms. Some are effective against a wide range of infectious agents and are known as "broad spectrum," whereas others are more specific. Unfortunately, for many infecting and disease-producing microorganisms (pathogens) there are as yet no effective antibiotics.

Vaccines, in contrast, are preparations of either dead organisms (or specific fractions thereof) or living attenuated microorganisms that may be administered to man or animals to stimulate their immunity to infection by the same or closely related organisms. The effectiveness of vaccines in controlling disease can vary widely. For example, smallpox vaccine, commonly prepared from the vaccinia lesions on the skin of inoculated calves or sheep or from the allantoic membranes of inoculated chick embryos, produces a high degree of immunity in vaccinated individuals and has been highly effective in controlling the disease. Measles vaccine, also prepared from an attenuated live virus, is likewise highly effective and produces a degree of immunity that is demonstrable by measurement of antibody titers, usually for a period of 8-10 years or more. These two vaccines produce immunity by infection with live virus and thus are long-acting. Influenza, on the other hand, may be prevented by the parenteral injection of vaccines of influenza viruses that have been grown in chick embryos and rendered noninfectious by formalin or ultraviolet irradiation. With such a dead vaccine, the duration of immunity rarely exceeds one year, and, since the causative agent (virus) of influenza is capable of mutating frequently, different vaccines usually must be prepared to combat each mutant strain.

Antibiotics and vaccines are both used in modern medicine to counter infectious diseases. Vaccines are used as prophylactic agents to increase immunity and to prevent subsequent infection with more virulent strains; antibiotics are employed most often as therapeutic agents to control disease after its onset, although they may also be used prophylactically against some organisms. There are, in addition, other antimicrobial drugs (not derived from growing microorganisms) which may be used for control of infection. These are generally used therapeutically, although at times, as with malaria, they may be used to prevent infection.

In developing countries, the capacity to produce many of these disease-fighting agents may only be realized further in the future than most other processes described in earlier chapters of this report. Local manufacture of drugs or vaccines requires considerable capital, a high degree of technology, and specially trained personnel; the use of these assets to produce antibiotics or vaccines locally must be viewed critically in relation to other needs for the same limited resources.

Economies of scale usually make it possible for industrialized countries to produce drugs more cheaply for large markets. Yet, in these industrialized countries, the major portion of research on drugs and vaccines is likely to relate directly to those diseases occurring within these industrialized areas, that is, not to those in the developing countries where the need is greatest.

Many diseases of developing countries, including the major tropical diseases, have attracted little attention and it may be that these diseases can best be attacked by a concerted research effort in the countries where they are endemic. Moreover, in view of the limited resources in these countries, sharing of effort on a regional basis may be desirable at times; but even so, the investment of time and money for this purpose in no way would be justifiable *until* other necessary elements of the public health system were adequate. These elements include:

- Public health statistics. Without good information on the principal causes of illness and death, health planning is ineffective and development of a health policy is impossible. A means of disseminating public health information is a prerequisite for public health programs.
- Public health programs. The source and mode of transmission of the target disease must be known. Effective programs may require control of nonhuman hosts or their environments. For example, in conjunction with drug administration, molluscicides may be required against snails for control of schistosomiasis, milk pasteurization for control of typhoid fever, or the destruction of mosquito-breeding areas for control of malaria.
- A health delivery system. An effective delivery system is a prerequisite to achieving health policy objectives. An effective disease-control program cannot be mounted without adequate storage facilities (generally low-temperature) and adequate means for drug distribution and administration.
- Knowledge of the nutritional status of the population and its relationship to disease. The interactions between an individual's nutritional state and

infection must be taken into account in designing an effective disease-control program.

This information on the health system will determine the kinds of vaccines or antibiotics required and whether their local manufacture might be economically feasible.

In this regard, modern drug development programs use increasingly sophisticated and sensitive analytical, chemical, and biochemical techniques. Drug intermediates are frequently extracted from plants or microorganisms and modified chemically to produce active substances. The distinction between microbiologically derived antibiotics and other antimicrobial drugs becomes less important as the criteria of efficacy and economics of production methods determine the extent to which microbes should be employed to synthesize a particular drug.

The World Health Organization (WHO) provides assistance through the establishment of integrated disease-control programs in most countries. The objectives of this program include assistance in:

- Developing a consistent approach to health policies;
- Assuring safety and efficacy of drugs;
- Achieving optimum utilization of drugs;
- Providing supplies of drugs;
- Exchange of data on drug experience among countries;
- Training of scientific personnel in the health field; and
- International collaboration in research and development on improved drugs.

Antibiotics and vaccines can be powerful tools for enhancing the health and well-being of afflicted populations. The virtual eradication of smallpox throughout the world is a fine example of the success of a coordinated disease-control program. With greater use of microbial processes and with worldwide collaboration other successes will follow.

Antibiotics

Probably no group of substances has been more thoroughly exploited than the antibiotics. Yet no antibiotics exist today for effective treatment of many of the diseases most prevalent in developing countries.

This is partly because of greater attention and resources being given to the diseases of industrialized countries; it is also because of the inherent nature of the diseases and their effects on human hosts. To be effective against a disease-producing organism the antibiotic must be able to inhibit its growth or its production of toxic substances without having a similar effect on surrounding host cells. This may be especially difficult when the organism is localized in a particular tissue (as in the case of tuberculosis or leprosy) where it is necessary to maintain a high concentration of the antibiotic over sufficient time without adverse effects on the human or animal host.

Further, there is often a need to grow the disease-producing organisms outside the human or animal body, so that experimentation with potential control agents will be possible. With some organisms, this has not yet been achieved. This inability to culture certain infecting organisms *in vitro* remains a major obstacle to their control.

Many antibiotic-producing microorganisms have been isolated from soil by a simple procedure: a soil sample containing millions of microorganisms is suspended in water, the suspension is diluted severalfold, and samples are transferred to Petri plates containing agar media of various nutritive compositions. The plates are incubated until growth of the microorganisms occurs in the form of individual colonies. If any of these colonies has produced an antibiotic, it probably will have diffused into the agar and it can be detected by spraying the agar with a suspension of bacteria susceptible to the antibiotic. The susceptible organisms form a solid lawn (growth) except where the antibiotic has diffused. Where the antibiotic exists, their growth will be arrested and a clear zone will surround the colony that has produced the antibiotic. Figure 9.1 illustrates zone formation for several different antibiotics.

Studies then can be carried out to determine the types of organisms inhibited by the antibiotic. The antibiotic-producing culture is grown in increasingly larger volumes (in Erlenmeyer flasks or fermentation vessels) after determining the conditions favoring maximum production of the desired active material. The antibiotic material may then be isolated and tested for potency and for its ability to produce side reactions or toxicity in animals. Finally it can be prepared in still larger quantities (Figure 9.2). Although basically

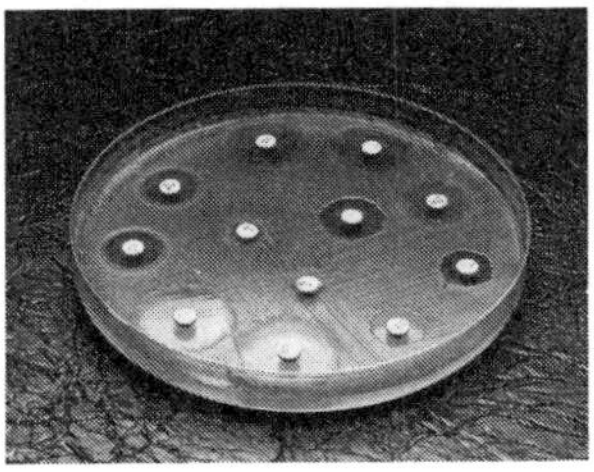

FIGURE 9.1 The effectiveness of several antibiotics against a single microorganism can be simultaneously determined. Discs containing a variety of agents are tested to enable selection (according to zone sizes) of those which might be used *in vivo*. (Photograph courtesy of BBL Microbiology Systems, Division of Becton, Dickinson and Company)

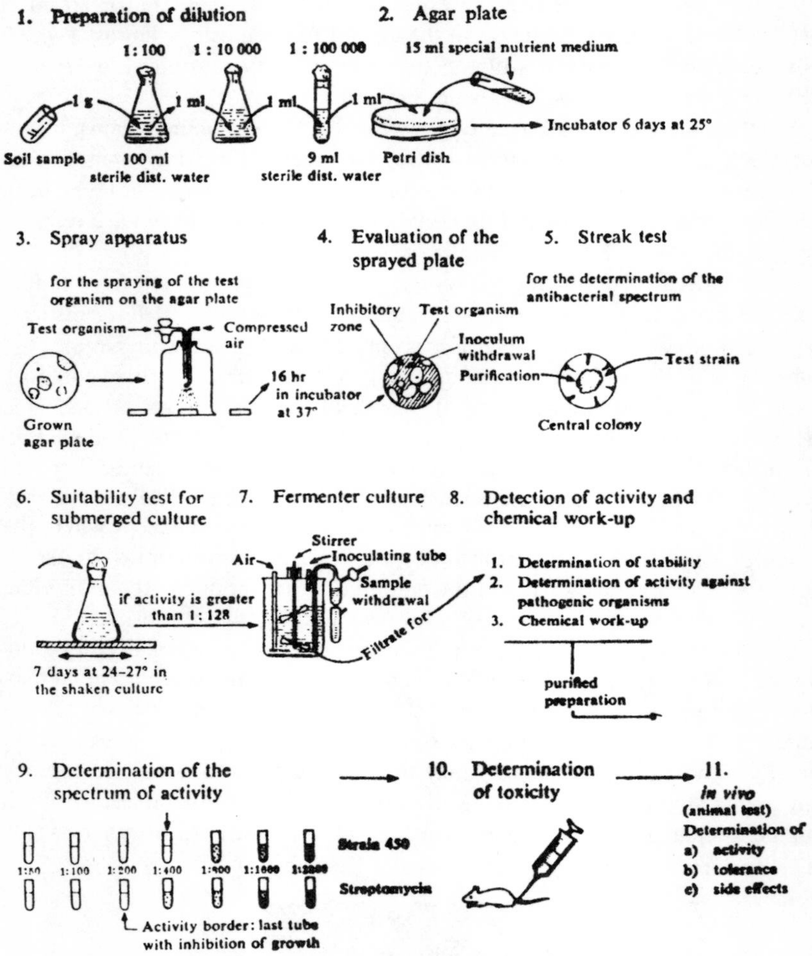

FIGURE 9.2 Antibiotic screening program. (Based on Reiner, 1977, p. 3.)

simple, an antibiotic-screening program may require months, sometimes years, of sustained work, teams of highly skilled personnel, and large investments in chemicals, laboratories, and fermentation and extraction equipment. Only rarely does this process yield a new antibiotic of real therapeutic and economic value.

Antibiotics may be distinguished on the basis of their spectrum of activity. Broad-spectrum antibiotics like chloramphenicol and the tetracyclines may affect many species of microorganisms and also unrelated organisms such as the rickettsiae, chlamydiae, and mycoplasma. Penicillin and streptomycin are examples of narrow-spectrum antibiotics that act against only a few bacterial species.

The major microbial sources of useful antibiotics are the actinomycetes (*Streptomyces*, *Nocardia*, *Micromonospora*) and molds or fungi (*Penicillium*, *Cephalosporium*).

Human Diseases

Table 9.1 lists major tropical diseases and indicates the availability of vaccines and antibiotics for their treatment. Much research remains to be done to arrest tropical diseases, particularly schistosomiasis, malaria, filariasis, trypanosomiasis (including African sleeping sickness and Chagas disease), onchocerciasis, and leishmaniasis, all common parasitic infections in the tropical zone. They are widespread, chronic, and may affect every member of a community. Although they are rare in industrialized countries in the temperate zones, the number of people suffering from their effects in the developing tropical regions runs into hundreds of millions. Their impact on individual well-being and productivity, in addition to physical suffering, has a profound effect on economic and social development.

Most therapeutically useful antibiotics exert their inhibitory effects only on bacteria. Since many of the tropical diseases are caused by protozoa, multicellular animal parasites, and viruses, other agents must be sought. Methods must be developed for cultivation of the parasites under laboratory conditions and for detection and assay of agents that can inhibit their growth.

Although the list of pathogens yet to be brought under control is lengthy, specific drugs and antibiotics discussed below have been developed for a number of widespread diseases.

Antifungal Agents Both superficial and systemic diseases result from infection of the skin and viscera by pathogenic fungi.

Typical skin infections, like chromomycosis and mycetoma, are caused by a diverse group of soil organisms. These infections develop slowly and are usually chronic. Mycetoma has been treated with penicillin and chromomycosis with amphotericin B, but surgical excision of small lesions is most effective. Better drugs are needed.

Histoplasmosis is a lung disease caused by inhaling spores from soil-borne fungi. The causative organism prefers soils contaminated by bat or bird droppings; caves and chicken coops are notorious sources of the infection. Amphotericin B has been used successfully in the treatment of histoplasmosis.

Blastomycosis is also a lung disease with secondary infections appearing on the skin. Amphotericin B has been used in the treatment of this disease.

Tinea favosa is a fungal infection of the skin and scalp commonly seen in children. Griseofulvin has been shown to produce good curative effects.

TABLE 9.1 Availability of Antibiotics, Chemicals, or Vaccines for Treatment or Prevention of Various Tropical Diseases (x = reported successful use; ? = possible use or testing)

Diseases	Antibiotics	Chemicals	Vaccines
Protozoan Diseases			
Amebiasis (Amebic dysentery)	x	Quinolines	
Leishmaniasis (Kala-azar)	?	Antimonials	
Malaria	?	Quinolines	
Trypanosomiasis (Sleeping sickness)		Amidines, Suramin	
Fungal Diseases			
Blastomycosis	x		
Chromomycosis	x		
Histoplasmosis (Darling's disease)	x		
Mucormycosis	x		
Mycetoma	x	Sulfas	
Rhinosporidiosis			
Sporotrichosis	x	Iodides	
Tinea favosa (Favus)	x		
Tinea imbricata (Tokelau)	x		
Bacterial Diseases			
Brucellosis (Undulant fever)	x		
Cholera	x		x
Cutaneous Diphtheria			x
Gonorrhea	x		
Leprosy	?	Sulfones	
Leptospirosis (Weil's disease)	x		?
Relapsing Fever	x	Arsenicals	
Shigellosis (Bacillary dysentery)	x		
Syphilis	x		
Tetanus			x
Tuberculosis	x	Isoniazides	x
Typhoid Fever	x		x
Yaws	x		
Rickettsial Diseases			
Boutonneuse Fever	x		
Epidemic Typhus	x		x
Murine Typhus	x		x
Scrub Typhus	x		

Mucormycosis is an infection of the intestines, lungs, and central nervous system, or skin. It is caused by one of a class of opportunistic fungi that generally do not infect the normal host but may cause disease in debilitated patients. For example, mucormycosis of the lung is associated with leukemia and mucormycosis of the intestines with malnutrition. In some instances, amphotericin B has been effective.

Antibacterial Agents Many drugs are available for the treatment and control of tuberculosis, a disease of bacterial etiology that is still highly prevalent

TABLE 9.1 Continued

Diseases	Antibiotics	Chemicals	Vaccines
Viral Diseases			
Bwamba Fever			
Dengue			?
Infectious Hepatitis			
Influenza A&B			x
Measles (Rubeola)			x
Poliomyelitis			x
Rabies			x
Rift Valley Fever			
Smallpox (Variola)	x		x
Trachoma		Sulfonamides	
West Nile Fever			
Yellow Fever			x
Zika Fever			
Helminth Diseases			
Ancylostomiasis (Hookworm)		Benzimidazoles	
Ascariasis (Roundworm)		Benzimidazoles	
Clonorchiasis (Oriental liver fluke)		Phenolics	
Dipetalonemiasis		Pyrazines	
Dipylidiasis		Salicylamides	
Dracunculiasis (Dragonworm)		Pyrazines	
Fascioliasis (Sheep liver fluke)		Phenolics	
Filariasis (Elephantiasis)		Pyrazines	
Gastrodisciasis		Benzimidazoles	
Gnathostomiasis (Creeping eruption)		Pyrazines	
Hydatid disease			
Hymenolepiasis (Dwarf tapeworm)		Salicylamides	
Loiasis (Eyeworm)		Pyrazines	
Onchocerciasis (River blindness)		Pyrazines	
Paragonimiasis (Oriental lung fluke)		Phenolics	
Schistosomiasis (Bilharziasis)		Nitroimidazoles	
Strongyloidiasis		Benzimidazoles	
Taeniasis (Tapeworm)		Salicylamides	
Trichinosis		Benzimidazoles	
Trichostrongyliasis		Benzimidazoles	
Trichuriasis (Whipworm)		Benzimidazoles	

in developing countries. When properly used, isoniazid and rifampin in combination with either streptomycin, *p*-aminosalicylic acid, or thiacetazone are highly effective therapeutic agents. Effective control of tuberculosis has been achieved, however, only in those countries where public health practices permit early detection of the disease by tuberculin testing and chest X-radiography, and by early treatment of all those who have had contact with those who are infected. Continuing control has been successful only where long-term drug administration can be assured.

Leprosy is a chronic infectious disease caused by *Mycobacterium leprae.* General treatment includes enhancement of personal and environmental hygiene and an ample, well-balanced diet. Although the sulfones (particularly

DDS, 4, 4′ - diamino diphenyl sulfone) have been the most effective class of drugs against leprosy, recent studies with the antibiotic rifampin at the WHO laboratories in Caracas have shown promise.

The widespread use of antibiotics in the treatment of bacterial infection has resulted in the emergence of numerous highly resistant pathogens. Drug resistance has developed through selection and transfer of genetic information (plasmids, extrachromosomal DNA) in microorganisms. Prevalence of drug-resistant strains makes it imperative that new drugs be developed for treatment of certain infectious diseases, particularly those caused by *Haemophilus influenzae*, *Neisseria gonorrhoeae*, and pathogenic *Enterobacteriaceae.*

Other Antimicrobial Agents Of the rickettsial diseases, epidemic typhus occurs in Africa and Europe. It is caused by *Rickettsia prowazekii* and can be controlled with chloramphenicol. Boutonneuse fever, a disease of the spotted fever group common in Africa, is caused by *R. conorii.* Chlortetracycline and chloramphenicol have produced favorable results in patients with this disease.

The protozoal disease amebiasis (amebic dysentery) is an infection by the ameba *Entamoeba histolytica.* Although amebiasis is not confined to the tropics, its incidence is determined in part by the level of sanitation in a given area. The majority of antibiotics showing activity against this ameba act indirectly through their bacteriostatic activity rather than directly against this parasite. However, two of them, oxytetracycline and fumagillin, appear to act directly as amebacides and at times have been found effective in the treatment of acute amebic dysentery.

Malaria may be the most serious widespread protozoal disease. The organisms, *Plasmodium* sp., are difficult to culture and are controlled only slowly by the tetracyclines. The disease is currently controlled by chemicals (mainly chloroquine derivatives) administered prophylactically and is treated chemotherapeutically by a variety of substances chemically related to quinine. WHO has cautioned against widespread use of antibiotics for treatment of malaria because of the risks of developing resistance in pathogenic bacteria, and recommends antibiotics only in areas where chloroquine-resistant strains of plasmodia are found.

Antitumor Agents The use of products of microorganisms has also been extended to the area of cancer chemotherapy.

Antibiotics with antitumor properties are produced by some microorganisms and a few of these may be used therapeutically (i.e., adriamycin, bleomycin, actinomycin D, mithramycin, mitomycin C).

The antitumor antibiotics are not a homogeneous group of compounds. Their antibacterial and antitumor activities are not correlated, but some correlation has been found between activities against transplantable tumors and human tumors.

Antihelminths The only effective therapeutic agents as yet developed against infestations in man and animals are chemical agents derived by other than microbial processes (see Table 9.1).

Animal Diseases

The most prevalent diseases of animals occurring in developing countries are trypanosomiasis (sleeping sickness), Newcastle disease, rinderpest, African swine fever, gastrointestinal parasitism, pasteurellosis, bovine pleuropneumonia, heartwater, sheep pox, babesiosis, brucellosis, hog cholera, and foot-and-mouth disease.

Trypanosomiasis occurs principally in Africa where the tsetse fly, by which it is spread, is found in 700 million hectares of land and prevents the raising of cattle. It is estimated that this vast area could support 125 million cattle and an equal number of sheep and goats. There are no drugs or vaccines for this parasitic disease, and development of effective ones is critically needed.

Gastrointestinal parasitism occurs worldwide in cattle, sheep, goats, swine, and poultry. A number of parasites are involved, although the helminths (roundworms) are the ones of prime importance. No vaccines are available, but there are effective drugs for control of some of these parasites.

Pasteurellosis also occurs worldwide but is important primarily in Africa and Asia. This bacterial disease affects cattle and buffalo and has a major impact on draft animals. Vaccines and antibiotics are effective against the disease, but delivery to the field in many countries presents difficulties.

Brucellosis occurs in naturally infected domestic animals in all parts of the world. Brucellosis is a serious cause of abortion in cattle, and to a lesser degree in sheep, goats, and swine. Vaccines and antibiotics are available for both prevention and cure of this bacterial disease.

Hog cholera is a highly contagious viral disease. Although there are no effective drugs, vaccines are available for control.

Foot-and-mouth disease is widespread in Europe, South America, the Middle East, and Mexico. Although vaccines may be used for control of this viral disease, they are type specific. To limit the spread of an outbreak, the enzootic strains must be determined and specific vaccines prepared for use against them. Control in many countries is through maintenance of quarantine along with supervised vaccination.

Newcastle disease occurs worldwide and affects poultry, a major protein source in developing countries. Although there are vaccines available for some forms of this viral disease, they are not effective in all field situations. There are no effective drugs.

Antibiotics as Growth Stimulants

In some geographic areas, antibiotics are now used extensively in animal feeds to promote weight increases of young animals. In developed countries, the use of such additives is economically important to farmers. When added to the feed of livestock and poultry in low concentrations (20–50 g per ton of feed), the animals are healthier, grow more rapidly, and reach marketable weight faster than those not fed antibiotics.

In the United States, six antibiotics are used extensively for growth-promoting effects on poultry, swine, cattle, and dairy calves: bacitracin, chlortetracycline, oxytetracycline, monensin, procaine penicillin, and tylosin. Improved growth rates and feed utilization are in the range of 2–15 percent for broilers, 2–13 percent for swine, 3–4 percent for beef cattle, and 10–30 percent for dairy calves. Production of these antibiotics by a low-level "non-sterile" manufacturing process (see below), if one could be developed, would significantly increase their usefulness as feed supplements by decreasing their cost.

The complication introduced by the appearance of mutant pathogenic microorganisms with multiple resistance to broad-spectrum antibiotics has led to consideration of banning of such feed additives in Great Britain and the United States. An antibiotic task force has questioned their use in feeds because of 1) development and dissemination of drug-resistant microorganisms, 2) increased shedding of salmonellae in animal dung, and 3) ingestion of antibiotic residues in human food, which may cause bacteriological and pharmacological hazards.

The possible spread of drug-resistant microorganisms through transfer of extrachromosomal genetic material (DNA) is now of international concern to scientists and governments. Because of this concern, the use of antibiotics as food preservatives has already been discontinued in the United States and England.

Plant Diseases

Hundreds of bacterial and fungal species and a few viruses that cause plant diseases are known to be suppressed by antibiotics. However, the use of antibiotics to control plant pathogens is carried out on a large scale only in Japan. Streptomycin, the tetracyclines, cycloheximide, and griseofulvin have been used most extensively, but all have serious drawbacks, including plant toxicity and high production costs. Blasticidin S and kasugamycin are widely used instead of mercurial fungicides to control the sheath blight of rice plants. These antibiotics are applied at very low concentrations, and the amount sprayed is 1–10 percent of that of conventional pesticides. This low concentration and the eventual biodegradation of the antibiotics lowers the possibility of environmental pollution. However, the emergence of resistant microorganisms reduces the attractiveness of the procedures.

Nonsterile Production of Antibiotics

Since the large-scale production of antibiotics requires considerable capital investment and includes steps (maintenance of cultures, fermentation) that must be carried out under scrupulously clean conditions, their manufacture in the developing countries should be considered only in exceptional situations. To set up low-level, low-cost production of carefully selected antibiotics for plant or animal disease control might be possible under less rigid conditions than those currently in use. The production of feed-quality chlortetracycline on cereal solids, which was developed in Czechoslovakia, is an example of the potential of this simplified technology.

Vaccines

There are many vaccines available for the prevention of infections in human beings. They differ considerably in their composition, effectiveness, and duration of protection. Some are live attenuated viruses; others consist of whole killed bacteria. Still other preparations consist of viral or bacterial components or of modified products of bacterial toxins (or toxoids). A list of diseases for which relatively effective vaccines or immunogenic agents are available is shown in Table 9.1. The characteristics and future possibilities for some of these agents are shown in Table 9.2.

Vital statistics obtainable from a number of countries suggest that utilization of many of the available vaccines is less than optimal. From the standpoint of public health and for the welfare of the population at large, it is important that nations establish effective programs for immunization against highly communicable infectious diseases. Such programs can be carried out only if the vaccines are available in adequate supply, with proper facilities provided for their storage and for protection of their potency, and only if the public can be educated about the benefits of immunization and a system of effectively delivering the vaccine can be developed and maintained. Identification of vaccine failures, especially where the use of live vaccines is concerned, is important in detecting flaws in the methods of vaccine storage or administration.

Certain vaccines offer a high degree of protection against the diseases for which they were developed. All children should be protected by immunization against measles, whooping cough, diphtheria, tetanus, and poliomyelitis. There has not been a naturally acquired case of smallpox reported in over a year anywhere in the world. If the disease is eradicated, then there will no longer be a need to vaccinate against it. Yellow fever vaccine has also proved highly effective and should be used in areas where the disease is endemic. Vaccines of the capsular polysaccharides of meningococcal Groups A and C

TABLE 9.2 Current and Future Vaccines and Protective Agents

Disease	Current Vaccine or Protective Agent	Possibilities in 5–10 Years
Cholera	Killed whole organism	1. Toxoid vaccine 2. Oral attenuated 3. Oral killed
Diphtheria	Toxoid-absorbed	Tetanus and diphtheria toxoids have been effectively used via the respiratory tract for booster immunization (concern exists for allergic reactions in the lung)
Tetanus	Toxoid-absorbed	
Pertussis	Killed bacteria or bacterial fraction	
Viral hepatitis		
Type A	Passive immune serum globulin (ISG)	Killed or attenuated vaccines
Type B	Passive immune serum globulin (ISG)	Specific high-titered ISG vaccine
Influenza	Egg-grown virus, formalin inactivated, highly purified by zonal ultracentrifugation	1. Live attenuated vaccine 2. Aerosol killed vaccine–fair protection, absence of side effects 3. Tissue culture-grown virus
Mumps	Vaccine	
Measles (rubeola)	Live attenuated virus, chick embryo or canine tissue-culture grown	
Plague	Formaldehyde-inactivated *Yersinia pestis*	
Poliomyelitis	Inactivated virus Attenuated virus monovalent or trivalent	
Rabies	Active: (1) B-propiolactone-inactivated virus grown in embryonated duck eggs (2) phenol-inactivated virus grown in rabbit brain Passive: equine hyperimmune serum	Live attenuated vaccine
Rubella	Vaccine	
Typhoid	Whole organism killed by several different techniques	Oral killed or attenuated vaccines have been tested and seem to be more effective with fewer side-effects
Typhus	Formaldehyde-inactivated *Rickettsia prowazekii* grown in embryonated eggs	
Yellow fever	Live attenuated virus prepared in chick embryo: Dakar strain or 17D strain	

Source: Robert H. Waldman. 1978. Immunization procedures. In *Clinical concerns of infectious diseases,* L. E. Cluff, and J. E. Johnson, eds. Baltimore: Williams & Wilkins. Co.

have been shown to be effective where cerebrospinal meningitis caused by these organisms is epidemic, and their administration is preferable to that of prophylactic antimicrobial drugs. To date, a vaccine effective against Group B meningococcal infection, although needed, has not been developed.

Parasitic and Venereal Diseases

A number of other widespread diseases still lack vaccines that will enable their eradication or control. These include the parasitic diseases endemic to many developing tropical countries and the venereal diseases, which are increasingly common throughout the world.

Parasitic Diseases Diseases caused by protozoa and helminths are among the major scourges of mankind. Malaria, trypanosomiasis, leishmaniasis, schistosomiasis, and filariasis afflict millions of people. If vaccination against these and other diseases caused by parasites is contemplated, it will first be necessary to cultivate the causative organisms and identify the significant cell components. The recent *in vitro* cultivation of malarial parasites in the United States and of the blood form of trypanosomes in Kenya may provide promising leads in the development of vaccines for these prevalent disorders.

Venereal Diseases Among the venereal diseases, gonorrhea is a major problem and in many geographic areas the etiological agent of the disease (*Neisseria gonorrhoeae*) has become resistant to the action of penicillin. There are significant gaps, moreover, in our knowledge of this organism and of the immune response to this infection in humans. Additional research is needed before a prophylactic vaccine can be developed. Meanwhile, antibiotics other than penicillin (for example, spectinomycin) may be effective. Syphilis is amenable to control by case- and contact-finding and by penicillin therapy. A vaccine effective against genital strains of *Herpes simplex* virus has distinct potential utility.

Respiratory Infections

In addition to diseases of childhood spread via the respiratory system (measles, rubella, mumps), there are many other infections transmitted in this fashion.

Pneumococcal Infections Of the respiratory infections of bacterial origin, those caused by pneumococci are most prevalent and are a problem in some nations where rapid urbanization is in progress. Polyvalent vaccines of pneumococcal capsular polysaccharides are currently available and should prevent the majority of pneumococcal pneumonias. A similar vaccine for preventing respiratory infection and meningitis in young children caused by *Haemophilus influenzae* Type B is under development, but the immune response of infants to this vaccine has been limited, as has the response to several other bacterial polysaccharides in children 12–18 months of age and younger. The need for research on the immunologic responsiveness to vaccines of this kind in children under 18 months is great.

Bacterial Otitis Media An infection of the middle ear, bacterial otitis media is prevalent in all societies. If effective vaccines for its prevention could be developed, the damage to hearing that often results from it could be eliminated.

Tuberculosis A vaccine of attenuated mycobacteria, BCG (Bacille Calmette Guérin strain), has been successfully used against tuberculosis in certain societies. Some current opinion supports the view that the effect is nonspecific and that an active public health program of case finding and treatment of each index case and its contacts is a more effective means of controlling tuberculosis.

Gastrointestinal Infections

Enteric or intestinal infections exact a terrible toll throughout most of the world. The World Health Organization estimates that approximately 70 million people are afflicted with significant diarrheal illness during each day of the year. Intestinal infections may be caused by a wide variety of bacteria, viruses, and protozoan parasites. They spare no age group, race, nation, or socioeconomic group. The young, especially infants, are particularly affected. In many countries diarrhea accounts for 25–50 percent of all infant deaths. Overall, enteric infection is the leading cause of mortality in most of the developing world.

Much can be done to eliminate diarrheal disease by sanitary measures, including the provision of good water supplies, sanitary disposal of sewage, and adequate cooking and refrigeration of foodstuffs. In areas where economic circumstances preclude such provision, vaccines offer a partially satisfactory means of preventing some gastrointestinal infection.

Various vaccines have been devised to prevent intestinal illness, and some have been in use since the latter part of the 19th century. Only in the last two decades, however, has their efficacy been properly evaluated by controlled trials. Unfortunately, these trials have shown that the available vaccines against typhoid fever and cholera are limited in their effectiveness, and there are no vaccines available for many other causes of diarrhea.

Typhoid Fever Long a scourge throughout the world, typhoid fever runs a protracted course, causing death in 10–20 percent of untreated victims and prolonged disability in its survivors. A recent outbreak in Mexico affected thousands of people, demonstrating the continuing threat of this disease.

Vaccines against typhoid fever have been in use since 1895. They consist of suspensions of killed typhoid bacilli. Side effects are frequent and include painful swelling at the injection site, fever, and malaise, all of which may persist for several days.

Because typhoid fever is generally contracted after ingestion of contaminated food or water, attempts have been made to stimulate intestinal immunity by the oral administration of killed typhoid bacilli. But even when these were given in twice the recommended dosage, the protective effect was only 30 percent. A recent attempt at oral immunization involved the use of attenuated typhoid bacilli. Given in multiple large doses to 155 adult volunteers, this live vaccine caused no side effects and protected 87 percent of individuals from subsequent illness.

Cholera Cholera, which ranks with typhoid as a global affliction, is known to have spread in pandemic fashion throughout the world on six occasions during the past two centuries. The seventh pandemic, which is currently affecting much of Asia, Africa, and the European countries bordering the Mediterranean, has been caused by a different type of organism, the so-called El Tor vibrio. The concern caused by the current appearance of the disease is great. Not only can cholera produce mortality rates of 50 percent and more in severely affected and untreated individuals, but it may create considerable panic and economic dislocation throughout a large geographic area.

As with typhoid fever, vaccines against cholera have been in use since the turn of the century. They have generally been composed of killed microorganisms administered by one or more injections. Although many claims have been made as to their effectiveness, field trials have demonstrated that vaccine efficacy at best was 76 percent during the first 6 months. Protection waned thereafter, and subsequent booster doses were required. These results were obtained in a population that had had frequent prior exposure to the etiological agent of cholera and presumably possessed considerable immunity prior to vaccination. The vaccine is generally given in mass fashion during fresh outbreaks, although it has never been shown to have prevented an epidemic.

Promising results have been obtained with a living attenuated mutant of the cholera organism given orally to adult volunteers in the United States. Approximately 60 percent of recipients, given one to four vaccine doses, were protected from illness after subsequent exposure.

The protective effect of this oral vaccine appeared to be at least equal to that of the injected vaccines tested in Bangladesh and elsewhere. Indeed, the efficacy of the oral preparation may be better in a true field situation, where the effective dose may prove to be substantially lower than that given to the volunteers of the U.S. study described above. Advantages of the oral product are that it avoids the painful side effects of injections and the skill and expense of administering them.

Present limitations to this oral vaccine are that it is live and produces small amounts of cholera toxin. Thus the possibility exists that it could revert to a virulent form. Neither this nor any other oral preparation has yet been field tested.

Escherichia coli So-called "enteropathogenic" serotypes of *E. coli* were first discovered in England in the 1940s. They have been considered to be associated primarily with epidemics of severe gastroenteritis in hospital nurseries. Today, many clinical laboratories have the facilities to readily identify these organisms.

It has been recognized only recently that other varieties of *E. coli*, termed "toxigenic" are probably far more important as causative agents of diarrheal illness. These organisms are similar to those causing cholera in that a toxin elaborated by the bacterium causes the illness, rather than the bacterium itself. They are probably responsible for a significant portion of the familiar traveler's diarrhea. Although epidemiological investigations are in their infancy, it seems quite likely that toxigenic *E. coli* play a substantial role in causing diarrhea throughout the world. Three other bacterial species, *Campylobacter fetus, Yersinia enterocolitica*, and *Vibrio parahemolyticus*, have been recognized recently as causes of diarrheal disease in man. No vaccine is available at present, nor is any mode of therapy definitely established.

Shigellosis Bacterial dysentery, or shigellosis, is an endemic problem in many areas of the world and frequently flares up in epidemic form. Striking examples include the large outbreaks in Central America and Bangladesh in recent years.

Treatment with specific antibiotics (chloramphenicol, ampicillin, or tetracycline) is usually successful, but shigella organisms possess a striking ability to develop resistance to antibiotic agents. An effective vaccine would be highly desirable, but none is commercially available.

Viral Diarrhea Various viruses have long been thought to be responsible for diarrheal illness. Recent studies have incriminated at least the reovirus and rotavirus. In some population groups, the latter agent is believed to be responsible for as much as 70 percent of diarrheal illness in infants. Other studies have found that young children rapidly acquire antibodies against these viruses, an indirect indication of the prevalence of infection.

Since work in this area is new, no vaccines are available. With successful cultivation and propagation of these viruses, however, it is reasonable to expect that an effective vaccine will be forthcoming.

Parasites There are many parasites that may cause significant diarrhea (*Giardia lamblia* and *Entamoeba histolytica* are excellent examples). There is no imminent promise of a successful vaccine against any of these.

References and Suggested Reading

Antibiotics

Beeson, P. B., and McDermott, W., eds. 1975. *Textbook of medicine*. Philadelphia-London-Toronto: W. B. Saunders Company.

Corcoran, J. W., and Hahn, F. E., eds. 1974. *Mechanism of action of antimicrobial and antitumor agents.* Antibiotic Series, Volume 3. New York: Springer-Verlag.

Goldberg, I. H.; Beerman, T. A.; and Poon, R. 1977. Antibiotics: nucleic acids as targets in chemotherapy. In *Cancer: a comprehensive treatise*, F. F. Becker, ed., Volume 6: *Chemotherapy*, pp. 427-456. New York: Plenum Press.

Kurylowicz, W., ed. 1976. *Antibiotics–a critical review*. Warsaw: Polish Medical Publishers, distributed in the United States of America and Canada by the American Society for Microbiology, Washington, D.C. 20006.

Lane, M. 1977. Chemotherapy of cancer. In *Cancer: diagnosis, treatment and prognosis,* 5th edition, J. A. Del Regato; H. J. Spjut; and J. Harlan, eds., pp. 105–130. St. Louis, Missouri: C. V. Mosby Company.

Maegraith, B. G., 1973. *One world*. London: Athlone Press, distributed in the United States of America by Humanities Press, Inc., Atlantic Heights, New Jersey.

_____. 1974. Tropical medicine: trends and progress. *Journal of Tropical Medicine and Hygiene* 77:4-7.

Reiner, R. 1977. Antibiotics. In *Methodicum chimicum*, Vol. 11: *Natural compounds*, Part 2: *Antibiotics, vitamines and hormones*, F. Korte and M. Goto, eds, pp. 2–68. New York: Academic Press.

Woodbine, M., ed. 1976. *Antibiotics and antibiosis in agriculture with special reference to synergism.* London: Butterworths.

Zahner, H., and Maas, W. K. 1972. *Biology of antibiotics.* New York: Springer-Verlag.

Zeigler, E.; Moody, W.; Hepler, P.; and Varela, F. 1977. Light-sensitive membrane potentials in onion guard cells. *Nature* 270:270-271.

Vaccines

American Academy of Pediatrics. 1974. *Report of the committee on infectious diseases.* 17th edition. Evanston, Illinois: American Academy of Pediatrics.

Ashcroft, M. T.; Singh, B.; Nicholson, C. C.; *et al.* 1967. A seven-year field trial of two typhoid vaccines in Guyana. *Lancet* 2:1056-1059.

Current status and prospects for improved and new bacterial vaccines. 1977. *Journal of Infectious Diseases* 136:Supplement.

Cvjetanovic, E., and Vermura, K. 1965. The present status of field and laboratory studies of typhoid and paratyphoid vaccines: with special reference to studies sponsored by the World Health Organization. *World Health Organization Bulletin* 32:29-36.

Dupont, H. L.; Hornick, R. B.; Snyder, M. J.; *et al.* 1972. Immunity in shigellosis. II. Protection induced by oral live vaccine or primary infection. *Journal of Infectious Diseases* 125:12-16.

Germanier, R. 1975. Effectiveness of vaccination against cholera and typhoid fever. *Monographs in Allergy* 9:217-230.

Gilman, R. H.; Hornick, R. B.; Woodward, W. E.; Dupont, H. L.; Snyder, M. J.; Levine, M. M.; and Libonati, J. P. 1977. Evaluation of a UDP-glucose-4-epimeraseless mutant of *Salmonella typhi* as a live oral vaccine. *Journal of Infectious Diseases* 136:717-723.

Gorbach, S. L., and Khurana, C. M. 1972. Toxigenic *Escherichia coli*: a cause of infantile diarrhea in Chicago. *New England Journal of Medicine* 287:791-795.
Hejfec, L. B. 1965. Results of the study of typhoid vaccines in four controlled field trials in the U.S.S.R. *World Health Organization Bulletin* 32:1-14.
Honda, T., and Finkelstein, R. 1979. Selection and characteristics of a *Vibrio cholerae* mutant lacking the A (ADP-Ribosylating) portion of the cholera entero-toxin. *Proceedings of the US National Academy of Sciences.* 76: 2052-2056.
Hornick, R. B., and Woodward, W. E. 1966. Appraisal of typhoid vaccine in experimentally infected human subjects. *Transactions of the American Clinical and Climatological Association* 78:70-78.
Mel, D. M.; Terzin, A. L.; and Vuksic, L. 1965. Studies on vaccination against bacillary dysentery. 3. Effective oral immunization against *Shigella flexneri* 2a in a field trial. *World Health Organization Bulletin* 32:647-655.
Mosely, W. H. 1969. The role of immunity in cholera. *Texas Reports on Biology and Medicine* 27:227–241.
National Academy of Sciences. 1979. *Pharmaceuticals for developing countries: conference proceedings of the division of international health of the institute of medicine.* Washington, D.C.: National Academy of Sciences.
Rotaviruses of man and animals: Editorial. 1975. *Lancet* 1:257.
Sack, R. B.; Hirschhorn, N.; Brownlee, I.; *et al.* 1975. *Entero*-toxigenic *Escherichia coli*-associated diarrheal disease in Apache children. *New England Journal of Medicine* 20:1041–1045.
Shore, E. G.; Dean, A. G.; Holik, K. J.; *et al.* 1974. Enterotoxin-producing *Escherichia coli* and diarrheal disease in adult travelers: a prospective study. *Journal of Infectious Diseases* 129:577–582.
Waldman, Robert H. 1978. Immunization procedures. In *Clinical concepts of infectious diseases*, L. E. Cluff, and J. E. Johnson, eds. Baltimore: Williams & Wilkins Co.
Woodward, W. E.; Gilman, R. H.; Hornick, R. B.; *et al.* 1976. Efficacy of a live oral cholera vaccine in human volunteers. *Developments in Biological Standardization* 33:108-112.
World Health Organization. 1972. *Oral enteric bacterial vaccines.* Technical Report Series, No. 500. Geneva: World Health Organization.
Yugoslav Typhoid Commission. 1964. A controlled field trial of the effectiveness of acetone-dried and inactivated and heat-phenol inactivated typhoid vaccines in Yugoslavia: Report. *World Health Organization Bulletin* 30:623–630.

Source of Cultures

Antibiotics and Vaccines

American Type Culture Collection, 12301 Parklawn Drive, Rockville, Maryland 20852, U.S.A.

Research Contacts

Antibiotics and Vaccines

Burton Pogell, School of Medicine, St. Louis University, 1402 South Grand Boulevard, St. Louis, Missouri 53104, U.S.A.

Oldrich K. Sebek, Infectious Disease Research Unit, The Upjohn Company, Kalamazoo, Michigan 49001, U.S.A.

William E. Woodward, Program in Infectious Diseases and Clinical Microbiology, University of Texas Health Science Center, P.O. Box 20708, Houston, Texas 77025, U.S.A.

Chapter 10

Pure Cultures for Microbial Processes

Mankind has utilized microbial fermentations to prepare foods and beverages for thousands of years. Two types of inocula have traditionally been used to initiate such fermentations: the natural flora associated with the products being fermented, such as yeasts on grapes in wine making, or a small amount of the previous fermented material containing the active microorganisms, as in yogurt culture. In both types of fermentations, the inocula usually consist of a mixture of microorganisms. Occasionally conditions may favor the growth of undesirable organisms normally present in small numbers, as may occur with acetic-acid-producing bacteria, contaminating a fermentation designed to produce an alcoholic beverage.

Techniques for producing a pure culture, that is, one containing a single type or strain of microorganism, were first developed by Robert Koch in the mid-19th century. These methods were immediately adopted by the microbiologists of the time, who were principally concerned with microorganisms as causes of disease. Such pure culture techniques led eventually to the definition and characterization of the bacteria that cause anthrax, tuberculosis, scarlet fever, and other major diseases. Similar procedures were developed for fungi, algae, and protozoa. The discovery of viruses came later, and because of their obligatory parasitism and submicroscopic size, they were much more difficult to characterize.

Pure culture techniques were in turn applied to commercial fermentations, which provided more consistent yields of the products desired. Such cultures have been used to make alcohol, yogurt, and citric and lactic acid and other useful products. The techniques also made possible the development of vaccines and antibiotics.

Development of Pure Culture Collections

Pure culture collections are important for a number of reasons. First they provide a source of reference to enable microbiologists to verify more easily the organisms with which they are working. They also provide a readily accessible source of cultures of known organisms and a means of preserving genetic resources of such organisms.

As pure cultures came into general use, they began to be collected on a systematic basis in a number of countries. The collections vary greatly in size, and some are quite specialized. In many cases, bacteria and protozoa are not kept in the same collection. There are a few exceptions, which will be noted below.

Pure culture collections have gained an important role as resources for authentic, reliable microbial cultures for both research and practical use. The need for international cooperation in establishing such culture collections has become increasingly evident with the development of important microbial biosynthetic processes. As a result, the International Federation of Type Culture Collections and the British Commonwealth Collection of Microorganisms were established in 1947. The International Association of Microbiological Societies approved the formation of a "Section on Culture Collections" in 1963, and in 1970 the Section was reorganized as the World Federation of Culture Collections (WFCC). National federations of culture collections exist in a number of countries, and at the urging of the Japanese federation, UNESCO brought together a group of culture collection specialists in Paris in 1966 to consider various problems relating to such collections. The training of culture collection personnel and the establishment of collections in developing nations were among the topics discussed. It became evident in the discussion that more information was needed on the location, content, and personnel of culture collections throughout the world.

Major Pure Culture Collections

A World Directory of Collections of Microorganisms was prepared in 1972 by S. M. Martin, of Canada, and V. B. D. Skerman, of Australia, with support from UNESCO, the World Health Organization, the Australian Commonwealth Scientific and Industrial Research Organization (CSIRO), and the Canadian National Research Council.

The directory provides a relatively complete list of collections throughout the world, but it will obviously require periodic updating. The location and nature of a few of the major collections are given in Table 10.1. The dates of publication of the most recent catalogues of cultures issued by these collections are shown in parentheses. It should be noted that most service collections charge a fee for providing cultures, in order to support the maintenance, characterization, and preservation of the cultures and to recover the costs of shipping them.

World Data Center and Microbiological Resource Centers

Although the importance of pure culture collections for the preservation of microbial germ plasm has long been recognized internationally, problems frequently arise when a comparison is made of the fermentative properties or other characteristics of microbial cultures listed under the same species name in different culture collections.

About 15 years ago V. B. D. Skerman, Professor of Microbiology, University of Queensland, Brisbane, Australia, began preparing a computerized list of the strains maintained by various institutions and their characteristics. This cataloging operation has been expanded over the years and is now known as the World Data Center (WDC). Recently, the WDC has taken on the task of documenting the characteristics of viruses on behalf of the International Committee on the Taxonomy of Viruses, as well as cataloging specialized collections such as those with strains of *Rhizobium* species for legume inoculation. Information concerning the WDC may be obtained from Professor Skerman.

Activities similar to those of the WDC have been carried on by other groups. For example, because of the importance of microorganisms in the production of antibiotics, the International Streptomyces Project was initiated in 1958. In this remarkable collaborative effort strains of important organisms were carefully studied and deposited in several of the culture collections (Table 10.1). Similar data for higher fungi are maintained in the Fungal Genetics Stock Center, Humboldt State University, Arcata, California.

In 1974 the concept of Microbiological Resource Centers (MIRCENs) was proposed to a group of microbiologists by M. K. Tolba, Executive Director, United Nations Environment Programme (UNEP). Financial support for putting this concept into action was provided by UNEP and UNESCO. One of the several aims of the MIRCENs is to provide the infrastructure for a network that will incorporate regional and interregional units geared to the management, distribution, and utilization of microbial gene pools. The first step in initiating the concept was the establishment of the WDC as a MIRCEN in close collaboration with the World Federation for Culture Collections. This MIRCEN serves as a pivotal unit for the formation of culture collections in developing countries and for providing data services to the centers acting in liaison with the WDC.

A regional MIRCEN at the Applied Scientific Research Corporation of Thailand in Bangkok serves microbiologists of Southeast Asia through the exchange of economically important microbial strains in the region, and by offering training and fellowship programs and promoting research on organisms relevant to the region. In the specialized area of microbial nitrogen fixation, similar MIRCENs have been developed in the Department of Soil Science and Botany, University of Nairobi, Kenya, and at the Instituto de Pesquisas Agropecuarias in Porto Alegre, Rio Grande do Sul, Brazil. The MIRCEN at Ain Shams University, Cairo, Egypt, serves the Middle East, and the one at the Central American Research Institute for Industry in Guatemala will function in the Central American region.

TABLE 10.1 A Representative List of Major Culture Collections

Australia

Fungi, yeasts, and actinomycetes (1973 catalogue)

Australian National Reference
Laboratory in Medical Mycology
The Institute of Medical Research
The Royal North Shore Hospital of Sydney
St. Leonards, N.S.W., Australia

Bulgaria

Pathogenic bacteria and fungi (1970 catalogue)

Bulgarian Type Culture Collection
Institute for State Control of Medical Preparations
Sofia, Bulgaria

Czechoslovakia

Bacteria, mycoplasmas, viruses, and fungi (1975 catalogue)

Czechoslovak Collection of Microorganisms
U.E. Purkyne University Frida OBRANCU
Miru 10, 66243
Brno, Czechoslovakia

France

Bacteria (1975 catalogue)

Collection of the Institut Pasteur
25, rue du Docteur Roux
Paris, France 75015

Germany

Bacteria, fungi, yeasts, and plant viruses (1974 catalogue)

Deutsche Sammlung von Mikroorganismen

Germany Continued

D-3400 Gottingen
Grisebachstrasse 8, West Germany

Hungary

Bacteria, fungi, and virus (1975 catalogue)

Microbiological Gene Bank
Microbiological Department Group of the Department of Food Technology and Microbiology
University of Horticulture
1064 Budapest, Izabella U. 46, Hungary

India

Bacteria and fungi (1971 catalogue)

Indian Type Culture Collection
Indian Agricultural Research Institute
New Delhi 12, India

Japan

Bacteria, fungi, viruses, bacteriophages, algae, protozoa, and rickettsia (1968 catalogue)

Japanese Federation of Culture Collections of Microorganisms
c/o Institute of Applied Microbiology
University of Tokyo, Bunkyo-ku
Tokyo, Japan

Fungi, yeasts, bacteria, and bacteriophages (1972 catalogue)

Institute for Fermentation
4-54 Juso-nishinocho
Higashiyodogawaku
Osaka, Japan

Bacteria and fungi (1971 catalogue)

Laboratory of Culture Collection of Microorganisms

Although the MIRCEN concept applies mainly to the less-developed countries, a technological MIRCEN has been established at the Karolinska Institute in Stockholm. This MIRCEN collaborates actively with the WDC in mapping metabolic characteristics of microorganisms and in helping other culture-collection personnel organize their specialized research projects.

Preservation Methods

A description of the maintenance of a small collection of microorganisms has been provided by Skerman (1977), including a method of scheduling the transfer of cultures. Skerman notes that, while a wide variety of culture

Japan Continued

Faculty of Agriculture
Hokkaide University
Sapporo, Japan

Actinomycetes (1976 catalogue)

Kaken Chemical Company, Ltd.
6-42, Jujodai 1-chome, Kita-ku
Tokyo 114, Japan

The Netherlands

Centraalbureau voor Schimmel Culture
Oosterstraat 1
Baarn, The Netherlands

United Kingdom

Algae and protozoa (1976 catalogue)

Culture Centre of Algae and Protozoa
Institute of Terrestrial Ecology
36 Storey's Way
Cambridge, CB3 ODT, England

Bacteria of industrial importance (1975 catalogue)

National Collection of Industrial Bacteria
Torry Research Station
PO Box 31, 135 Abbey Road
Aberdeen AB9 SDG, Scotland

Bacteria of medical and veterinary importance (1972 catalogue)

National Collection of Type Cultures
Central Public Health Laboratory
Colindale Avenue
London NW9 5HT, England

Bacteria pathogenic for plants (1971 list)

National Collection of Plant Pathogenic Bacteria

United Kingdom Continued

Plant Pathology Laboratory
Hatching Green
Harpenden, Hertfordshire
AL5 2BD, England

Fungi (other than animal pathogens and wood-rotting fungi) (1975 catalogue)

Collection of Fungus Cultures
Commonwealth Mycological Institute
Ferry Lane
Kew, Surrey TW9 3AF, England

United States of America

Bacteria, fungi, algae, protozoa, bacteriophages, cell cultures, viruses, antiserum, rickettsiae and chlamyciae (1978 and 1979 catalogues)

American Type Culture Collection
12301 Parklawn Drive
Rockville, MD 20852, U.S.A.

Bacteria and fungi (no catalogue)

Northern Regional Research Laboratory
U.S. Department of Agriculture, Science and Education Administration
1815 North University Street
Peoria, IL 61604, U.S.A.

World List

Rhizobium Collections (1973 catalogue)

International Biological Program
World Catalogue of Rhizobium Collections
7 Marylebone Road
London N.W. 1, England

Source: S. M. Martin, and V. B. D. Sherman, eds. 1972. *World directory of collections of cultures of microorganisms.* New York: John Wiley and Sons.

media may be required for growth of organisms held in a collection, it is desirable that the number of media be kept to a minimum. In the collection that Skerman describes, involving about 1,200 strains of bacteria and fungi plus a few algae, 68 percent of the organisms could be grown on two kinds of media. Yet for the entire collection 57 different types of media were required.

For a much larger collection (approximately 25,000 strains), the American Type Culture Collection (ATCC) stores over 1,000 culture media and these are listed in the ATCC Catalogue of Strains published in 1978. As with the smaller collection designed by Skerman, many of the bacterial strains can be grown on relatively few media. Nevertheless, a wide variety of media are needed for the more fastidious organisms in a collection, making a culture collection somewhat expensive to maintain, depending on the types and number of cultures in it.

In the early days of culture collections, the cultures were maintained by serial transfer, that is, from culture grown in laboratory tubes or dishes to fresh medium. This method maintains the viability of a colony of microorganisms, but it is frequently ineffective in maintaining genetic integrity and ensuring that the important biosynthetic characteristics will not be lost or modified. Thus a strain of a microorganism developed to yield high levels of an antibiotic may gradually lose that important capability during continuous transfer in the laboratory, despite the ability of the microorganisms to multiply. Ironically, the ability to grow may often be enhanced as the loss of the ability to produce the desired compound occurs. This problem led microbiologists to seek other means of maintaining cultures.

Some of the methods that have been developed are quite simple and have proved useful for many strains of microorganisms.

They include drying the culture on sterile sand or soil, sterile filter paper strips, plastic spheres, or glass beads. Regardless of the method used, however, extensive laboratory studies of every highly developed strain are necessary to assure against loss of any economically important biosynthetic characteristic.

The development of freeze-drying (lyophilizing) procedures during the past 30–40 years represented a large step forward in preserving cultures. Lyophilization involves freezing a culture at very low temperatures (about $-65^\circ C$) in a mixture of dry ice and alcohol, after which the water is removed by sublimation from the solid state under a high vacuum.

Although freeze-drying will significantly stabilize the characteristics of many types of microorganisms, not all species will survive lyophilization. With continued improvements, however, a larger number and variety of microorganisms are surviving the process.

Even when an organism survives freeze-drying, the freeze-dried culture must be stored under controlled conditions. Some results of studies carried out at the ATCC using relatively hardy organisms are given in Table 10.2.

Even those organisms listed in Table 10.2 that were freeze-dried and stored for 11 months in a refrigerator at a temperature of $4^\circ C$ showed a drop in viable count to approximately half that of the original culture, while the same cultures stored at room temperature ($22^\circ C$) dropped to less than 1 percent of the original count. The advantage of the freeze-dried method, despite the loss, is that transfer is not necessary and the culture can be kept stored for years. Nevertheless, while viable cultures could be retrieved from either group, the question remains whether the survivors will retain the biosynthetic capability that is important. Therefore, where proper facilities are available, even freeze-dried cultures should be stored at as low a temperature as the laboratory can provide.

For many years dry ice was used to lower the temperature of heavily insulated boxes to -65°C. In recent years mechanical (electrical) refrigerators capable of maintaining temperatures as low as -75°C have been developed and are widely employed in modern laboratories.

TABLE 10.2 Survival of Freeze-Dried Cultures

Organism	Bacterial Counts × 10^{-6} Original Count	Storage Temperature Count after 11 months 4°C	22°C
Streptococcus faecalis	620	310	9.0
Pseudomonas aeruginosa	200	120	7.6
Escherichia coli	680	290	1.2
Enterobacter aerogenes	980	440	3.6
Average	620	290	5.3

Source: *American Type Culture Collection, Catalogue of Strains, I.* 1978. 13th edition. Rockville, Maryland: American Type Culture Collection.

The ultimate in present-day refrigerators are units cooled by liquid nitrogen. Storage temperatures as low as -196°C can be maintained. Liquid nitrogen storage units are excellent for maintaining almost all types of microorganisms, including algae, protozoa, and even mammalian tissues, in viable form. Since this type of equipment is not likely to be available in smaller collections, many smaller laboratories arrange to store key strains under liquid nitrogen in the larger culture collections.

In addition to the storage of cultures, culture collections are often responsible for research and education in culture maintenance, storage and characterization. Taxonomic studies are invaluable to culture collections since the material held in a collection must be properly identified and classified. The culture collection is a most appropriate location for taxonomic research, with many major contributions to scientific knowledge made in connection with culture collections.

Culture collections exist in many places apart from the major collection centers. Many organizations using microorganisms in agriculture and industry maintain small collections of organisms for their particular purposes. In developing countries, cultures, and the microbiologists who maintain and use them, may represent important resources that are not fully appreciated or utilized for national development objectives. Brought together, they could at a minimum serve as an expert source of advice and insight for development authorities into alternative ways in which microorganisms can be exploited for development objectives, such as those described in this report, in the context of local resources and constraints.

Mixed Microbial Cultures

The preoccupation over many decades with pure culture techniques is giving way to a second look at mixed microbial cultures. It has been clearly shown, for example, that *Chlorella*, among the green algae, can be cultivated effectively under nonsterile conditions. Bacteria are present, to be sure, but appropriately compounded nutrient media permit a growth pattern favoring *Chlorella* and prevent bacterial overgrowth.

Likewise, many foodstuffs customarily used in less-developed countries contain substantial numbers of mixed species of microorganisms. Mixed cultures of lactic acid bacteria are prominent in fermented foods derived from milk. Cheeses, curds, and cakes of various descriptions for human consumption have evolved in many parts of the world and are found even among primitive societies.

The subject of mixed-culture microbial technology is a fascinating one, and major portions of some international meetings are now devoted to this subject. The Symposium on Indigenous Fermented Foods presented as part of the Fifth International Conference on Global Impacts of Applied Microbiology (Bangkok, November 1977) covered many processes of this nature.

Patenting of Processes Involving Microorganisms

In many countries it is possible to obtain patents for products and processes involving microorganisms. To file for such a patent, it is usually necessary to deposit the microorganism(s) involved in a culture collection recognized for the purpose by the local patent authority. Both of the collections listed in Table 10.1 for the United States are recognized by the U.S. Patent Office for this purpose. (The Northern Regional Research Laboratory, however, will not accept pathogenic organisms, and in general restricts its collection to bacteria and fungi.)

In case of multinational filing of patent applications, it is necessary to deposit the cultures in each of the countries in which a patent is sought, except in those cases where countries have reached international agreements recognizing each other's depositories. Since this has been both troublesome and somewhat costly, the World Intellectual Property Organization (WIPO), which deals with international agreements on patents and copyrights, sought to develop an international treaty to make it possible to recognize the deposition of such organisms in a single depository as fulfillment of this requirement for all signatories to the treaty. Such a treaty was finally completed in Budapest, Hungary, in May 1977, and a number of the major nations have already signed it.

Although it may take some time before this treaty is activated, it would be well for microbiologists in all nations interested in seeking patents to be aware that it exists. For further information, inquiry can be made to the Secretary General, WIPO, Geneva, Switzerland.

References and Suggested Reading

American Type Culture Collection, Catalogue of Strains, I. 1978. 13th edition. Rockville, Maryland: American Type Culture Collection.

Colwell, R. R. 1975. *The role of culture collections in the era of molecular biology.* Washington, D.C.: American Society for Microbiology.

Martin, S. M., and Skerman, V. B. D., eds. 1972. *World directory of collections of cultures of microorganisms.* New York: John Wiley and Sons.

Skerman, V. B. D. 1977. The organization of a small general culture collection. In *Proceedings of the International Conference on Culture Collections–II, July 15-20, 1973, São Paulo, Brazil.* A. F. Pestana de Castro, E. J. Da Silva, V. B. D. Skerman, and W. W. Leveritt, eds., pp. 20-40. Bowen Hills, Queensland, Australia: Courier-Mail.

Steinkraus, K. H., ed. 1977. Papers presented at the Symposium on indigenous fermented foods, Fifth International Conference on Global Impacts of Applied Microbiology, November 21–27, 1977, Bangkok, Thailand. (Will be published under title *Handbook of tropical indigenous fermented foods.*) 21 *U.S. Code* 111.

Research Contacts

P. Atthasampunna, Thailand Institute of Scientific and Technological Research, Bangken, Bangkok 9, Thailand.

J. R. Jardim Freire, IPAGRO, Caixa Postal 776, 90000 Porto Alegre, Rio Grande do Sul, Brazil.

C.-G. Hedén, Karolinska Institutet, Solnavagen 1, S-104 01 Stockholm 60, Sweden.

S. O. Keya, University of Nairobi, P.O. Box 30197, Nairobi, Kenya.

V. B. D. Skerman, University of Queensland, St. Lucia, Brisbane, Queensland 4067, Australia.

Chapter 11

Future Perspectives in Microbiology

Certain major turning points in history have resulted from scientific breakthroughs, but the advances from the new knowledge have not been restricted to science. Rather they have also helped solve philosophical puzzles, changed economies, and often improved the quality of life.

Science and technology are being called on today to help mankind alter a pattern of life that has been lavish in the use of finite natural resources and turn to one more dependent upon renewable substances. Two significant kinds of renewable resources are those that depend on photosynthesis and those that take advantage of the useful and beneficial activities or properties of microorganisms.

In previous chapters of this report, a few well-known examples were cited to illustrate the impact that microbiology has had on human welfare. But what about the future? In this brief chapter, mention will be made of additional areas in microbiology that may—if certain developments occur—lead to great economic and social benefits as well as contributing to fundamental knowledge. Development of this potential need not be restricted to highly developed industrialized countries; in fact, some advances in microbiology may more easily come from less-developed regions of the world.

New basic techniques are being discovered and old ones improved in biology, biochemistry and chemistry. Some of these techniques are precise and relatively easy to perform. Others are more complex and require expensive instruments and facilities not available to all scientists. Still others depend upon microbiological techniques such as animal and plant tissue culture research. Plant tissue culture, for example, may lead to improved varieties of plants by enabling scientists to select mutants (both in haploid and diploid lines) and to study protoplast fusion, regeneration of whole plants, and other plant functions.

Much of our recent knowledge in genetics and molecular biology that is leading to so-called genetic engineering has been cradled in microbiology. There is little doubt that basic and applied research with microorganisms in these fields will continue to provide valuable information and technology of economic value. This knowledge can be used for improvements on commercial fermentation processes, in agriculture, in the pharmaceutical industry for production of improved bacterial and viral vaccines, and possibly in making substances to correct metabolic defects in human beings. For example, bacteria were recently engineered to make insulin by transplanting into them the gene from rat cells that carries the instructions to synthesize insulin. Much research remains to determine whether such insulin will function in human diabetics, if the bacteria can be implanted in the human intestine, and if they will then continue to produce the hormone. In a similar manner, a strain of *Escherichia coli* has been used to produce in the laboratory the human hormone somatostatin, which is normally formed in the hypothalamus at the base of the brain.

Research on plasmids formed by the bacterium *Agrobacterium tumefaciens* is giving us much basic information on how tumors are produced in plants. This knowledge may be helpful in other types of cancer research; other similar ways of using microorganisms for cancer study are in experimental stages.

Certain bacteria (*Halobacterium halobium*) contain in their cell membranes a purple protein pigment closely related to the visual pigment of vertebrates. Indications are that the vertebrate purple protein constitutes an essential link in the signal chain of the visual process in animals and human beings. Speculation is that research on the bacterial purple pigment will help explain the mechanism of how animals see.

Some microscopic marine green algae (*Dunaliella* species) grow in waters of high salt content (for example, the Red Sea and the Dead Sea), where they produce large quantities of glycerol. Possible commercial extraction of the glycerol from these algae is contemplated.

Microorganisms are known to produce a wide variety of metabolic products; in fact, over 5,000 metabolites have been identified and some 500 enzymes described. Most scientists believe these metabolites and enzymes represent only a fraction of the total existing in nature. It seems reasonable to assume that some of these substances may be useful and have economic value. In fact, interesting possibilities exist that some of these substances have pharmacological potential. For example, Japanese scientists have isolated from culture filtrates of actinomycetes (*Streptomyces testaceus*) a substance called pepstatin, possessing strong antipepsin activity. The substance is being used to analyze the role of pepsin in stomach or duodenal ulcers. Certain microorganisms produce antitumor compounds, which show promise for future production by fermentation and eventually for therapeutic use. A species in the bacterial genus *Nocardia*, for example, produces potent compounds called ansamitocins, which are active antitumor substances.

Culture filtrates of a fungus (*Fusarium oxysporum*) contain a compound identified as fusaric acid, which inhibits the enzyme dopamine-β-hydroxylase. This enzyme appears to be related in some way to Parkinson's disease in human beings. Also, oral administration of fusaric acid causes experimental animals and human beings to become sensitive to alcohol. Unpleasant side effects often result when persons eat mushrooms (*Coprinus atramentarius*) and drink alcoholic beverages. Scientists have recently discovered the basic anti-alcohol compound (coprin) in the mushroom, and it may become effective in the treatment of alcohol addiction. Further study of these relationships may contribute to our knowledge of Parkinsonism and alcoholism, and in turn have profound social, medical, and economic significance.

Similar examples can be cited of microbes producing substances that lower hypertension and destroy cholesterol in the blood, or serve as anti-inflammatory agents, neuromuscular blocking compounds, or other useful pharmacological agents.

Agriculture can also benefit from microbial research. One illustration among many is the unique product produced by bacteria (*Pseudomonas abikonensis*, *P. fianii*) that inhibits the growth of the bacterium (*Xanthomonas citri*) responsible for cankers on citrus-fruit trees and the fungus (*Piricularia oryzae*) that causes blast in rice.

The discovery that biologically active substances can be fixed artificially to insoluble polymers (such as membranes and particles), which act as supports or carriers, has greatly advanced certain areas of science and technology. Useful microbial enzymes that are rapidly inactivated by heat can be stabilized by attachment to inert polymeric supports, and in other cases these so-called "insolubilized microbial enzymes" can be used in nonaqueous environments. Whole bacterial cells can also be immobilized inside polyacrylamide beads and used for a variety of purposes. The possibilities seem limitless for the use of certain microbial cells and their products.

The transfer of microbial DNA to plant cells in nature appears to be of considerable importance in causing plant diseases and economic losses. This area of plant pathology and microbiology deserves more attention. For instance, crown gall in plants is initiated during the first few days of infection by the causative bacterium (*Agrobacterium tumefaciens*). But once the plant cells are transformed and have produced a gall, the living bacteria are no longer necessary to maintain the tumorous state. Tumor cells free of bacteria can be isolated from diseased plants and cultured *in vitro* by usual tissue-culture methods. Such cultured cells of crown gall proliferate as tumors when grafted onto suitable host plants, and in some cases even pass on the transformed characteristics to the healthy host cells. Study of these characteristics may be important to the understanding of mechanisms that control—or fail to control—orderly cell multiplication.

New hope for discovering chemical substances to cure virus diseases has arisen with the successful use of adenine arabinoside to treat herpes encephalitis, a virus disease that destroys brain cells. Similar viruses cause fever blisters, genital herpes, and other diseases. This advance may be comparable to the discovery of penicillin.

The possibility of using the unicellular alga *Cosmarium turpinni* as a protein supplement for animal feeds holds promise. When this alga is cultivated in the laboratory in the presence of cellulysin (from *Trichoderma reesei*) it forms only protoplasts (cells without rigid cell walls) either in light or in the dark. This obviates having to break cell walls to release cellular proteins, which is one of the problems associated with the use of single-cell proteins for livestock and poultry feeds.

Interest in geomicrobiology has been growing in recent years, with the development of new insights into the role of microbes in a number of geological processes. Microbes are now recognized as important geologic agents, playing a role in such geologic processes as mineral formation, mineral degradation, sedimentation, weathering, and geochemical cycling. From a human standpoint, these processes may either be beneficial or harmful, depending on the context.

Beneficial effects include microbial extraction by solubilization (leaching) of commercially useful substances. This enables metals like cobalt, copper, lead, zinc or uranium (see Chapter 6) to be separated from low-grade ores from which they cannot be economically extracted by more conventional methods of milling and flotation. Beneficial effects may also include the microbial genesis of sulfur from sulfate or of methane from organic residues in natural environments, immobilization or volatilization of polluting toxic elements such as arsenic or mercury, the microbial desulfurization of coal, the microbial removal of methane from coal mines, and the use of aliphatic hydrocarbon-utilizing bacteria in prospecting for petroleum deposits.

Harmful effects may be the microbial genesis of acid mine-drainage from microbial pyrite oxidation in bituminous coal seams, occurring after exposure to air and moisture during mining; the release of toxic substances such as antimony or arsenic from naturally occurring minerals into the environment; or the microbial weathering of building stone such as limestone, leading to defacement or structural weakness.

Discoveries of previously unknown microbial interactions with inorganics, like the deposition of manganese in nodules and crusts on the ocean floor, are continuing and will provide further insights into geological processes and are likely to yield many new practical applications of microbes for economic benefit.

For instance, two useful microbial processes are being tested for obtaining petroleum products from oil shale and tar sands. Carbonates in shale decrease permeability and hinder extraction of the oil. By applying certain bacteria that produce acids from shale components, the carbonates are dissolved in the shale matrix, increasing porosity and facilitating the removal of oil products. Many difficulties exist in the extraction of oil from the tar sands. A novel approach for obtaining hydrocarbons from such sands has been described whereby microbes adsorb and emulsify the oil, thus aiding in conventional processing methods.

The search for life on other planets has been largely unsuccessful, but it has provided new techniques for the identification of microbes and their products using automated and miniature apparatus. This kind of apparatus is finding important application in other fields of biology. For example, so-called pyrolysis-spectrometry techniques can distinguish between healthy and diseased or abnormal tissues in the body. High hopes exist for using these techniques to reduce the time required to identify genetic defects in fetal cells obtained by amniocentesis. The search for microbes on other planets may thus be responsible for important spin-offs that may have considerable significance in the future.

From these brief descriptions of microbial processes, it is apparent that the science of microbiology has reached a point where it can make real contributions to improving the welfare of mankind. The main question now is whether ingenious and well-informed microbiologists and bioengineers have the vision—and the ability—to convince the public that the beneficial activities of the microbial world can be exploited for human good.

References and Suggested Reading

Anonymous. 1977. Proteins from synthetic genes. *Nature* 270:202.

Borowitzka, L. J., *et al.* 1977. The salt relations of *Dunaliella*. Further observations on glycerol production and its regulation. *Archiv fuer Mikrobiologie* 1113:131–138.

Brady, R. O., *et al.* 1974. Replacement therapy for inherited enzyme deficiency. Use of purified glucocerebrocidase in Gaucher's disease. *New England Journal of Medicine 291:989.*

Brierley, C. L. 1978. Bacterial leaching. *CRC Critical Reviews in Microbiology* 6:207-262.

Da Silva, E. J., Olembo, R., and Burgers, A. 1978. Integrated microbial technology for developing countries: springboard for economic progress. *Impact of Science on Society* 28:159-182.

Dimmung, W. 1977. Feedstocks for large-scale fermentation processes. In *Microbial energy conversion,* H. G. Schlegel and B. Barnea, eds. Oxford: Pergamon Press.

Hedén, C.-G. 1977. Enzyme engineering and the anatomy of equilibrium technology. *Quarterly Review of Biophysics* 10:113-135.

Henderson, R. 1978. The purple membrane of halobacteria. In *Relations between structure and function in the prokaryotic cell: 28th Symposium of the Society for General Microbiology*, University of Southhampton, April 1978, R. Y. Stanier, H. J. Rogers, and J. B. Ward, eds., pp. 225-231. Cambridge-New York: Cambridge University Press.

Litchfield, J. M. 1977. Single-cell proteins. *Food Technology* 31:5:175-179.

Prave, P. 1977. Utilization of microbes—modern developments in bacteriological technology. *Angewandte Chemie* (Eng. ed.) 16:205-213.

Snyder, H. E. 1970. Microbial sources of protein. *Advances in Food Research* 18:85-140.

Stanton, W. R., and Da Silva, E. J., eds. 1978. *GIAM V. Global impacts of applied microbiology. State of the art: GIAM and its relevance to developing countries.* pp. 323. Kuala Lumpur: University of Malaysia Press.

Tannenbaum, S., and Wang, D., eds. 1975. *Single-cell protein II.* Cambridge: MIT Press.

Viking Mission to Mars. *Science* 193(4255):759-815.

Waslein, C. I., *et al.* 1969. Human intolerance to bacteria as food. *Nature* 211:84-85.

Appendix

Regulations for Packaging and Shipping Viable Microbial Cultures

Most organisms used in industrial or biosynthetic processes are not known to cause animal or plant diseases, but even those that can cause disease have not been ignored in their application to such processes. In addition, disease-control programs may require shipments of disease-causing microbe cultures for identification or testing. Consequently, packaging of microbial cultures to be shipped either domestically or internationally has been regulated in many countries.

In the United States, for example, cultures must be packaged and shipped in accordance with the Code of Federal Regulations 49, CAB82, Tariff 6-D. International shipments must also comply with International Air Transport Association Regulation 736, Packaging Note 695. The appropriate portion of the Code of Federal Regulations can be purchased from the Superintendent of Documents, U.S. Government Printing Office, Washington, D.C. 20402, U.S.A. The International Air Transport Association Regulations may be purchased from the IATA, P.O.Box 160, Geneva, Switzerland.

If laboratories in developing nations want to ship cultures to the United States or to develop regulations in their own countries, it would be helpful to become acquainted with these rules.

Regulations for packaging and shipping of cultures vary greatly according to the country. Some countries (for instance, New Zealand, Canada, and Australia) require permits before any culture may be shipped into the country. The United States requires that export licenses be obtained from the Department of Commerce before bacteria, fungi, algae, protozoa, and viruses may be shipped out of the country.

Cultures classified as etiologic agents (defined as a microorganism or its toxin that causes or may cause human disease) must be properly packaged to withstand leakage of contents, shocks, pressure changes, and other conditions incident to ordinary handling in transportation. The package must be marked, including an etiologic agent/biomedical material label as specified by the Public Health Service of the U.S. Department of Health, Education, and Welfare. In addition, properly prepared forms for restricted articles must accompany the shipment. Certain microbe strains also require permits from the U.S. Department of Agriculture and/or the U.S. Public Health Service before shipment can be made.

Importation of cultures into the United States is regulated by the U.S. Department of Agriculture (USDA) and the Public Health Service.

It has been USDA policy not to permit the importation (except to Plum Island Animal Disease Center) of pathogens that cause the following diseases: rinderpest, foot-and-mouth disease, African swine fever, hog cholera, swine vesicular disease, African horse sickness, Rift Valley fever, Teschen, Nairobi sheep disease, lumpy skin disease, louping ill, bovine infectious petechial fever, Newcastle disease (Asiatic strains), sheep pox, camel pox, goat pox, ephemeral fever, vesicular exanthema, Borna disease, Wesselsbron disease, and a variety of organisms of lesser importance.

The shipment of cultures to other countries, as noted above, is similarly controlled by the local laws in those countries. Therefore, it is essential for culture collection personnel to know what permits may be required to transmit cultures to the countries of scientists requesting their cultures.

WILLIAM A. W. KREBS, Vice President, Arthur D. Little, Inc., Cambridge, Massachusetts (member through December 1977)

THOMAS F. MALONE, Foreign Secretary, National Academy of Sciences, *ex-officio*

FRANÇOIS MERGEN, Pinchot Professor of Forestry, School of Forestry and Environmental Studies, Yale University, New Haven, Connecticut

FREDERICK T. MOORE, Economic Advisor, International Bank for Reconstruction and Development, Washington, D.C. (member through December 1978)

W. HENRY MOSLEY, Director, Cholera Research Laboratory, Dacca, Bangladesh, and Associate, Johns Hopkins University, Baltimore, Maryland (member through December 1978)

RODNEY W. NICHOLS, Vice President, Rockefeller University, New York, New York

DANIEL A. OKUN, Department of Environmental Sciences and Engineering, School of Public Health, University of North Carolina, Chapel Hill, North Carolina

E. RAY PARISER, Senior Research Scientist, Department of Nutrition and Food Science, Massachusetts Institute of Technology, Cambridge, Massachusetts

JOSEPH PETTIT, President, Georgia Institute of Technology, Atlanta, Georgia (member through December 1977)

JOSEPH B. PLATT, President, Claremont University Center, Claremont, California (member through December 1977)

HUGH POPENOE, Director, International Programs in Agriculture, University of Florida, Gainesville, Florida

JAMES BRIAN QUINN, Amos Tuck School of Business Administration, Dartmouth College, Hanover, New Hampshire

PRISCILLA C. REINING, Director, Project on Desertification, International Division, American Association for the Advancement of Science, Washington, D.C.

RALPH W. RICHARDSON, JR., Department of Horticulture, Pennsylvania State University, University Park, Pennsylvania

FREDERICK SEITZ, President Emeritus, Rockefeller University, New York, New York

H. GUYFORD STEVER, Consultant, Washington, D.C.

VICTOR RABINOWITCH, Director

MICHAEL G. C. McDONALD DOW, Deputy Director

JOHN G. HURLEY, Deputy Director

Advisory Committee on Technology Innovation

GEORGE BUGLIARELLO, President, Polytechnic Institute of New York, Brooklyn, New York (Chairman through December 1978)

HUGH POPENOE, Director, International Programs in Agriculture, University of Florida, Gainesville, Florida, *Chairman*

Members

HAROLD DREGNE, Director, International Center for Arid and Semi-Arid Land Studies, Texas Tech University, Lubbock, Texas

ANDREW HAY, President, Calvert-Peat, Inc., New York, New York

CYRUS McKELL, Institute of Land Rehabilitation, Utah State University, Logan, Utah

FRANÇOIS MERGEN, Pinchot Professor of Forestry, School of Forestry and Environmental Studies, Yale University, New Haven, Connecticut

E. RAY PARISER, Senior Research Scientist, Department of Nutrition and Food Science, Massachusetts Institute of Technology, Cambridge, Massachusetts (member through 1977)

CHARLES A. ROSEN, Staff Scientist, Stanford Research Institute, Menlo Park, California (member through 1977)

VIRGINIA WALBOT, Assistant Professor, Department of Biology, Washington University, St. Louis, Missouri (member through 1977)

Board on Science and Technology for International Development

Commission on International Relations

National Academy of Sciences–National Research Council

2101 Constitution Avenue, Washington, D.C. 20418, USA

Advisory Studies and Special Reports

Reports published by the Board on Science and Technology for International Development are sponsored in most instances by the U.S. Agency for International Development and are intended for free distribution primarily to readers in developing countries. A limited number of copies is available for distribution on a courtesy basis to readers in the United States and other industrialized countries who have institutional affiliation with government, education, or research and who have professional interest in the subject areas treated by the reports.

Single copies of published reports listed below are available free while the supplies last. Requests should be made on your organization's letterhead. Other interested readers may buy the reports listed here from the National Technical Information Service (NTIS) whose address appears below.

5. **The Role of U.S. Engineering Schools in Development Assistance.** 1976. 30 pp. Examines opportunities and constraints facing U.S. engineering schools in mobilizing their resources to aid developing countries. NTIS Accession No. PB 262-055. $4.50

7. **U.S. International Firms and R,D & E in Developing Countries.** 1973. 92 pp. Discusses aims and interests of international firms and developing-country hosts and suggests that differences could be mitigated by sustained efforts by the firms to strengthen local R,D & E capabilities. NTIS Accession No. PB 222-787. $6.00

8. **Ferrocement: Applications in Developing Countries.** 1973. 89 pp. Assesses state of the art and cites applications of particular interest to developing countries—boat-building, construction, food and water storage facilities, etc. NTIS Accession No. PB 220-825. $6.50

14. **More Water for Arid Lands: Promising Technologies and Research Opportunities.** 1974. 153 pp. Outlines little-known but promising technolgies to supply and conserve water in arid areas. NTIS Accession No. PB 239-472. $8.00 (French-language edition is available from Office of Science and Technology, Development Support Bureau, Agency for International Development, Washington, D.C. 20523 or through NTIS, Accession No. PB 274-612. $8.00.)

16. **Underexploited Tropical Plants with Promising Economic Value.** 1975. 187 pp. Describes 36 little-known tropical plants that, with research, could become important cash and food crops in the future. Includes cereals, roots and tubers, vegetables, fruits, oilseeds, forage plants, and others. NTIS Accession No. PB 251-656. $9.00

17. **The Winged Bean: A High Protein Crop for the Tropics.** 1975. 43 pp. Describes a neglected tropical legume from Southeast Asia and Papua New Guinea that appears to have promise for combatting malnutrition worldwide. NTIS Accession No. PB 243-442. $4.50.

18. **Energy for Rural Development: Renewable Resources and Alternative Technologies for Developing Countries.** 1976. 305 pp. Examines energy technologies with power capabilities of 10-100 kilowatts at village or rural level in terms of short- and intermediate-term availability. Identifies specific research and development efforts needed to make intermediate-term applications feasible in areas offering realistic promise. NTIS Accession No. PB 260-606. $11.75. (French-language edition is available from Office of Energy, Development Support Bureau, Agency for International Development, Washington, D.C. 20523.)

19. **Methane Generation from Human, Animal, and Agricultural Wastes.** 1977. 131 pp. Discusses means by which natural process of anerobic fermentation can be controlled by man for his benefit, and how the methane generated can be used as a fuel. NTIS Accession No. PB 276-469. $7.25.

21. **Making Aquatic Weeds Useful. Some Perspectives for Developing Countries.** 1976. 175 pp. Describes ways to exploit aquatic weeds for grazing, and by harvesting and processing for use as compost, animal feed, pulp, paper, and fuel. Also describes utilization for sewage and industrial wastewater treatment. Examines certain plants with potential for aquaculture. NTIS Accession No. PB 265-161. $9.00.

22. **Guayule: An Alternative Source of Natural Rubber.** 1977. 80 pp. Describes a little-known bush that grows wild in deserts of North America and produces a rubber virtually identical with that from the rubber tree. Recommends funding for guayule development. NTIS Accession No. PB 264-170. $6.00

23. **Resource Sensing from Space: Prospects for Developing Countries.** 1977. 203 pp. An examination of current and prospective applications of interest to the LDCs, certain implications for long-term governance of a remote sensing system, and desirable technical cooperation initiatives to diffuse user capabilities. NTIS Accession No. PB 264-171. $9.25.

25. Tropical Leguemes: Resources for the Future. 1979. 331 pp. Describes plants of the family Leguminosae, including root crops, pulses, fruits, forages, timber and wood products, ornamentals, and others. NTIS Accession No. PB 298-423. $12.00.

26. Leucaena: Promising Forage and Tree Crop for the Tropics. 1977. 118 pp. Describes *Leucaena leucocephala*, a little known Mexican plant with vigorously growing, bushy types that produce nutritious forage and organic fertilizer as well as tree types that produce timber, firewood, and pulp and paper. The plant is also useful for revegetating hillslopes and providing firebreaks, shade, and city beautification. NTIS Accession No. PB 268-124. $7.25.

28. Microbial Processes: Promising Technologies for Developing Countries. 1979. 200 pp.

29. Postharvest Food Losses in Developing Countries. 1978. 202 pp. Assesses potential and limitations of food loss reduction efforts; summarizes existing work and information about losses of major food crops and fish; discusses economic and social factors involved; identifies major areas of need; and suggests policy and program options for developing countries and technical assistance agencies. NTIS Accession No. PB 290-421. $9.25.

30. U.S. Science and Technology for Development: Contributions to the UN Conference. 1978. 226 pp. Serves the U.S. Department of State as a major background document for the U.S. national paper, 1979 United Nations Conference on Science and Technology for Development. Includes an overview section plus five substantive sections as follows: 1) industrialization; 2) health, nutrition, and population; 3) food, climate, soil, and water; 4) energy, natural resources and environment; and 5) urbanization, transportation, and communication.

Related Publications

Other reports (prepared in cooperation with BOSTID) available from the above address are:

An International Centre for Manatee Research. 1975. 34 pp. Describes the use of the manatee, a large, almost extinct, marine mammal, to clear aquatic weeds from canals. Proposes a research laboratory to develop manatee reproduction and husbandry. Published by the National Science Research Council of Guyana. NTIS Accession No. PB 240-244. $4.50.

Natural Products for Sri Lanka's Future. 1975. 53 pp. Report of a 1975 workshop with the National Science Council of Sri Lanka. Identifies neglected and unconventional plant products that can significantly contribute to Sri Lanka's economic development. Published by National Science Council of Sri Lanka. NTIS Accession No. PB 251-520. $5.25.

Workshop on Solar Energy for the Villages of Tanzania. 1978. 167 pp. Report of a workshop with the Tanzania National Scientific Research Council, Dar es Salaam, Tanzania. Reviews state-of-the-art of small-scale solar energy devices, and suggests short- and long-range projects using them in villages. Published by Tanzania National Scientific Research Council. NTIS Accession No. PB 282-941. $9.00.

Out-of-Print Publications

The following out-of-print BOSTID reports are available **only** from the National Technical Information Service unless otherwise noted. To order, send report title, NTIS Accession Number, and amount indicated. (Note: Prices are current for April 1979 and are subject to change without notice.) Pay by NTIS Deposit Account, check, money order, or American Express account. U.S. orders without prepayment are billed within 15 days; a $5.00 charge is added. Prices for foreign buyers are double the prices indicated below, and payment in full must be enclosed. Send order to:

National Technical Information Service
Springfield, Virginia 22161, USA

1. East Pakistan Land and Water Development as Related to Agriculture. January 1971. 67 pp. Reviews World Bank proposed action program in land and water management. NTIS Accession No. PB 203-328. $5.25.

2. The International Development Institute. July 1971. 57 pp. Endorses concept of new science-based technical assistance agency as successor to AID; examines its character, purposes, and functions. NTIS Accession No. PB 203-331. $5.25.

3. Solar Energy in Developing Countries: Perspectives and Prospects. March 1972. 49 pp. Assesses state of art, identifies promising areas for R & D, and proposes multipurpose regional energy research institute for developing world. NTIS Accession No. PB 208-550. $5.25.

4. Scientific and Technical Information for Developing Countries. April 1972. 80 pp. Examines problem of developing world's access to scientific and technical information sources, provides rationale for assistance in this field, and suggests programs for strengthening information infrastructure and promoting information transfer. NTIS Accession No. PB 210-107. $6.00.

6. Research Management and Technical Entrepreneurship: A U.S. Role in Improving Skills in Developing Countries. 1973. 40 pp. Recommends initiation of a systematic program and indicates priority elements. NTIS Accession No. PB 225-129. $4.50.

9. Mosquito Control: Some Perspectives for Developing Countries. 1973. 63 pp. Examines biological control alternatives to conventional pesticides; evaluates state of knowledge and research potential of several approaches NTIS Accession No. PB 224-749. $6.00.

10. Food Science in Developing Countries: A Selection of Unsolved Problems. 1974. 81 pp. Describes 42 unsolved technical problems with background information, possible approaches to a solution, and information sources. NTIS Accession No. PB 235-410. $6.00.

11. Aquatic Weed Management: Some Perspectives for Guyana. 1973. 44 pp. Report of workshop with the National Science Research Council of Guyana describes new methods of aquatic weed control suitable for tropical developing countries. NTIS Accession No. PB 228-660. $5.25.

12. Roofing in Developing Countries: Research for New Technologies. 1974. 74 pp. Emphasizes the need for research on low cost roofs, particularly using materials available in developing countries. NTIS Accession No. PB 234-503. $6.00.

13. Meeting the Challenge of Industrialization: A Feasibility Study for an International Industrialization Institute. 1973. 133 pp. Advances concept of an independent, interdisciplinary research institute to illuminate new policy options confronting all nations. NTIS Accession No. PB 228-348. $7.25.

15. International Development Programs of the Office of the Foreign Secretary, by Harrison Brown and Theresa Tellez. 1973. 68 pp. History and analysis, 1963-1973; lists staff/participants and publications. NTIS Accession No. PB 230-543. $5.25.

20. Systems Analysis and Operations Research: A Tool for Policy and Program Planning for Developing Countries. 1976. 98 pp. Examines utility and limitations of SA/OR methodology for developing country application and means for acquiring indigenous capabilities. NTIS Accession No. PB 251-639. $6.50.

24. Appropriate Technologies for Developing Countires. 1977. 140 pp. Examines fundamental issues and inter-relationships among economic, political and social factors relating to choice of technologies in developing countries. Discusses criteria of appropriateness and suggests policies for improving technical decisions. Available from Office of Publications, National Academy of Sciences, 2101 Constitution Ave., N.W., Washington, D.C. 20418 USA; please enclose payment of $6.25.

Other out-of-print reports (prepared in cooperation with BOSTID) available from the National Technical Information Service are:

Products from Jojoba: A Promising New Crop for Arid Lands. 1975. 30 pp. Describes the chemistry of the oil obtained from the North American desert shrub *Simmondsia chinensis*. NTIS Accession No. PB 253-126. $4.50.

Aquatic Weed Management: Some prospects for the Sudan and the Nile Basin. 1975. 57 pp. Report of a 1975 workshop with the Sudanese National Council for Research. Suggests modern and innovative methods for managing the water hyacinth. Published by National Council for Research–Agricultural Research Council of Sudan. NTIS Accession No. PB 259-990. $5.25.

Ferrocement, a Versatile Construction Material: Its Increasing Use in Asia. 1976. 106 pp. Report of a 1974 workshop with the Asian Institute of Technology, Bangkok, Thailand. Surveys applications of ferrocement technology in Asia and the Pacific Islands. Includes construction of grain silos, water tanks, roofs, and boats. Published by Asian Institute of Technology. NTIS Accession No. PB 261-818. $6.50.

International Consultation on Ipil-Ipil Research. 1978. 1972 pp. Report of a 1976 conference sponsored with the Philippine Council for Agriculture and Resources Research, Los Baños, Laguna, Philippines. Contains background papers and workshop session summary reports on ipil-ipil (*Leucaena* spp.). (Companion volume to report no. 26 above.) NTIS Accession No. PB 280-161. $8.00.

Reports in Preparation (working titles)

BOSTID will fill requests for single copies of reports in preparation upon publication as outlined at the beginning of this section.

27. Firewood Crops: Shrub and Tree Species for Energy Production.
31. Food, Fuel, and Fertilizer from Organic Wastes.
32. The Water Buffalo: Its Potential for Development.
33. The Potential for Alcohol Fuels in Developing Countries.
34. Revegetating the Range: Selected Research and Development Opportunities.
35. Aerial Seeding of Forests.

ORDER FORM

While the limited supply lasts, a free copy of ***Microbial Processes: Promising Technologies for Developing Countries*** *will be sent to institutionally affiliated recipients (in government, education or research) upon written request on your organization's letterhead or by submission of the form below. Please indicate on the labels the names, titles, and addresses of qualified recipients and their institutions who would be interested to have this report.*

Please return this form to

Commission on International Relations (JH 215)
National Academy of Sciences–National Research Council
2101 Constitution Avenue
Washington, D.C. 20418, USA

28

28

28

28

28

28